Teradata® Database Index Essentials

Alison M. Torres

First Edition

Colors of God's Gifts, LLC

www.TeradataIndexes.com

Teradata® Database Index Essentials

©Alison M. Torres, 2011
Colors of God's Gifts, LLC

All rights reserved. No part of the contents of this book may be reproduced or transmitted in any form or by any means, including recording or by any information storage and retrieval system without written permission from the author.

First Printing 2011

ISBN: 978-0-9839154-0-9

The information in this document is provided on an "as-is" basis, without warranty of any kind. The author will not be liable for any indirect, direct, special, incidental, or consequential damages, including lost profits or savings. Information in this document may contain technical inaccuracies or typographical errors. The author has attempted to provide accurate and complete information in the creation of this book. All attempts have been made to verify information presented in this publication, the Author assumes no responsibility for errors, omissions, or contrary interpretations of the subject matter herein. This book is not intended for use as a source of legal, business, or any other advice. All readers are advised to seek services of competent professionals in legal, business, and other matters. This book is not published by, on behalf of, or under authority or direction of Teradata Corporation or its affiliates (Teradata). All opinions expressed and statements made by the author and by any other individual employed by or otherwise associated with Teradata are made by those people in their individual capacities, not as representatives of Teradata, and are not endorsed, adopted, or authorized by Teradata.

Trademarks

The following names are registered names and/or trademarks, and are used throughout the book: Teradata, BYNET, and the Teradata logo are trademarks or registered trademarks of Teradata Corporation and/or its affiliates in the U.S. and worldwide. The "Teradata" logo is used by permission of Teradata Corporation; Teradata retains all rights in the logo and its use. Linux is a registered trademark of Linus Torvalds. In addition to these product names, all brand and product names in this manual are trademarks of their respective holders in the United States and/or other countries.

Printed by Sky Marketing and Publishing
Rochester, NY
2011

Teradata® Database Index Essentials

Published by:

Sky Publishing and Marketing

1387 Fairport Road

Suite 800

Fairport, NY 14450

To order more copies of this book or to inquire about other products and services by Alison Torres contact:

http://TeradataIndexes.com

Alison M. Torres

Teradata® Database Index Essentials

Dedication

The gift of encouragement often comes from the least likely or most unexpected places!

I met Marcia and Walter Friedberg while on vacation with my Mom over thirty years ago. It was an odd feeling to be recognized by people I didn't know, especially in a foreign country. I never thought that encounter would change my career path nor have such an amazing impact on my life.

The Friedbergs "adopted" me into their family and have always wanted nothing but the best for me. They have mentored, encouraged, loved and fed me chicken soup for many years.

I remember asking Mr. Friedberg what I should call him, he responded, "Those who don't know me, call me Walter. Those who know me, call me Walt. Those who love me, call me Wally."

To my family - Mãe, Dad, Gary, Glen, Kara, Lace, Ryan, London, Ming, Asher, Lachlan, Lisa, Cindy, and Dave…I love you all and you'll have to wait your turn.

This book is for Wally! Everyone should have a Wally in their life.

Acknowledgments

Every book takes a great deal of time, dedication, and effort to put together. Most authors find that they really cannot do it all on their own.

I would like to acknowledge and thank everyone with whom I've worked over the years and have helped me develop my expertise, which significantly enhanced the value of this book. I would particularly like to acknowledge my two esteemed colleagues, Larry Higa and Larry Carter, for their insights and content contributions.

To Ann Kruglak, we had different plans for my visit, but this book is so much better because of the time we spent editing, not to mention the fun we had. Thank you.

To my book production angels, Rich Selby, my publisher, my editor, Kit Birmingham, and Rob Kalnitz, my graphic artist, for their invaluable advice in the publishing arena. Thank you, we really made it happen!

What Other People Are Saying About Teradata® Database Index Essentials

Alison Torres is one of the Teradata Education Network's most popular presenters on its live webcast series, and is definitely the most prolific, with many more topics recorded than any other presenter!

Working with Alison is wonderful; she is an expert in so many areas, she presents information in a clear and concise format, and she is so much fun to listen to!

She understands the issues and problems customers have completely, is able to take complex technical suggestions and break them down into understandable language and processes, and has fun passing the information on!

Elise M. Zob
Teradata Corporation
Teradata Education Network Program Manager,
Teradata Training

*****

Alison proposes us a concise, clear and very practical book on Teradata Database Indexes. Today I had the chance to have a peek at Alison's book on Teradata Indexes. Whatever index flavour you are looking for, it's in! UPI, NUPI, USI, NUSI, NoPI, PPI, MLPPI and all versions of JI, they are all there. This is a concise, clear and complete book, a perfect reference for DBA and SQL coder.

Don't miss the chance to get your signed copy at Partners. Mine is already ordered.

Thumbs up!

Patrice Bérubé
Solution Architect
Teradata US

Foreword

Los Angeles 1988:

Teradata User Group Conference (long before it became the Partners Conference). The small hotel meeting room was full of geeky hands-on IT folks, the earliest users of the Teradata technology. A young woman sat in the front row of the hotel meeting room looking a bit out of place among the hard core techies. She listened with interest and understanding, scribbling notes in her notebook, as I was presenting a very dry and detailed technical topic on the internals of Teradata. She quickly came to the front of the room when I finished my talk, eager to ask me a list of questions from her notes. Thus began my long and wonderful friendship with Alison.

Two decades later, Alison is still just as enthusiastic about Teradata as she was then. She is the one on stage teaching Teradata to anyone and everyone, a mission which has taken her to the far corners of the globe and put her in front of audiences of every possible description. Her gift combines her extraordinarily deep understanding of Teradata with her energy to bring easy understanding to what could be intensely tedious topics. She makes detailed technology into a story that delivers the knowledge and makes Teradata transparent while having fun and keeping the audience engaged. Alison now brings that same energy and clarity to her book on Teradata Indexes. In this volume Larry Carter, a long time Teradata educator, and Larry Higa, the grand wizard of Teradata performance optimization, contribute to her vast font of knowledge. Combined, they bring more than 70 years of experience in the subject to these pages. No need to wait to get on their incredibly crowded calendars, you can benefit from their knowledge on the topic anytime, anywhere.

So why Indexes anyway? The power of Teradata is its scalability, parallelism and complex query optimizer, why not focus on those? Because Alison knows that indexing is the primary tool a Teradata user has available to significantly impact performance. Done well, indexes can provide fast response times to questions and optimize the resource utilization of even the heaviest workloads. Done poorly or not at all, this can lead to widespread disappointment with the investment in a company's data warehouse. It is an area that is fundamentally different than on other databases. If techniques learned on other products are applied on Teradata, the result is often significant waste of resources and sub-optimal delivery of access to the users of the system.

Anyone working on Teradata for the first time, especially those with lots of experience and training on other platforms should read this book. And if you have been working on Teradata for a while, you will still find new opportunities to improve your system utilization and your users' experience. It is a great reference book for anyone who creates a table on Teradata, needs to optimize a new workload, or just wants to get more out of their current investment. I am willing to bet that even the most veteran Teradata user will find something of value in this book. Read this book, then keep it close by your desk as a reference to refer to when a difficult workload or performance opportunity presents itself.

Thank you Alison! Thank you for sharing your knowledge of Teradata in a way that all users can take advantage of any time. And thank you for being a wonderful friend.

Todd Walter
Chief Technologist
Teradata Americas

Preface

The purpose of this book is to provide a reference about Teradata Database indexes, not just what they are, but how they work and how to use them, sometimes in unconventional ways.

The content will be most useful for database administrators (DBAs) and application developers, in a Teradata Database environment.

Teradata Database Indexes serve two major purposes:
- Access to data
- Distribution of data

Indexes provide the greatest benefits for accessing table data. Choosing the right index makes data access more selective and cuts down on physical I/O, creating a huge cost savings. Indexes help get data into memory more efficiently, which can reduce the cost of processing answer sets. Indexes allow access to specific data rows and can eliminate accessing a lot of excess data rows. There are instances where Secondary or Join Indexes can be used to satisfy queries more efficiently than base table access would be.

Index Definition:

Aside from using a Primary Index to locate rows on AMPs in Teradata Database's parallel environment, indexes are data structures that can be used by the Teradata query optimizer to greatly improve table access performance.

There are situations where full-table scans are a reasonable access method and Teradata Database does handle them quite effectively. However, by using an index, the system can avoid performing full-table scans, especially when retrieving a relatively small number of rows from a very large table.

We will explore Teradata Database Indexes: what they are, how they work, and when to use them.

Table of Figures

Figure 1:	Teradata Node - Simplified	
Figure 2:	Teradata SMP (Symmetric Multi-Processor) Node	
Figure 3:	Teradata MPP (Massively Parallel Processing) Nodes	
Figure 4:	Primary Index Attributes	
Figure 5:	Primary Index RowHash Determines AMP Location for Storage	
Figure 6:	Partitioned Primary Indexes: How They Work	
Figure 7:	PI Range Query	
Figure 8:	Full Table Scan on AMP with PI	
Figure 9:	Partition Access on AMP with PPI	
Figure 10:	Partitioning with Range_N	
Figure 11:	Partition Inserts	
Figure 12:	Partitioning Error Message	
Figure 13:	Access Using Partitioned Data	
Figure 14:	Access Using Non-Partitioned Data	
Figure 15:	Access Using a PPI Table via the Primary Index with No Partition Specified	
Figure 16:	Accessing a NPPI Table via the Primary Index	
Figure 17:	Various PPI Enhancements	
Figure 18:	Create Time-Based PPI Table	
Figure 19:	Alter Time-Based PPI Table	
Figure 20:	Alter Effect on Time-Based PPI Table	
Figure 21:	Multi-Level Partitioned Primary Index Syntax	
Figure 22:	Sales for 2 Years With No Partitioning	
Figure 23:	Week 6 Sales Only and Single Level Partitioning	

Figure 24:	Week 6 Sales for District 25 and Multi-Level Partitioning
Figure 25:	CREATE Table Statement – Unique Partitioned Primary Index with three levels
Figure 26:	Sample Row Header for Sales Table with Three Levels of Partitioning
Figure 27:	Partition Claim Table by ClaimDate and StateID
Figure 28:	Access a Multi-Level Partitioned Primary Index
Figure 29:	Partition # Formula
Figure 30:	Collation and Case Sensitivity Considerations
Figure 31:	Claim Table with One Level of Partitioning
Figure 32:	Queries that will Benefit from Partitioning
Figure 33:	Claim Table with Three Levels of Partitioning
Figure 34:	Partition Elimination Query
Figure 35:	Multi-Level Partition Elimination Query
Figure 36:	Secondary Index Rules
Figure 37:	Create a Unique Secondary Index
Figure 38:	Access via a USI
Figure 39:	Create a Non-Unique Secondary Index
Figure 40:	Access via a NUSI
Figure 41:	NUSI ORDER BY Options
Figure 42:	NUSI Covering Criteria
Figure 43:	Three Column NUSI
Figure 44:	Value Ordered NUSI
Figure 45:	Range Constrained Query
Figure 46:	NUSI Index Covering
Figure 47:	Create a USI on a PPI Table

Figure 48:	Select on USI Column
Figure 49:	Create a USI on a Non-Unique PPI Table
Figure 50:	Select on NUSI Column
Figure 51:	Create a NUSI on a Non-Unique PPI Table
Figure 52:	Customer Table
Figure 53:	Orders Table
Figure 54:	Customers with Orders
Figure 55:	Compressed Multi-Table Join Index Syntax
Figure 56:	Compressed Multi-Table Join Index Query Results
Figure 57:	Non-Compressed Multi-Table Join Index Syntax
Figure 58:	Non-Compressed Multi-Table Join Index Query Results
Figure 59:	Compressed Join Index Storage
Figure 60:	Non-Compressed Join Index Storage
Figure 61:	Query to Determine Space Usage
Figure 62:	Join Index Space Usage Comparison
Figure 63:	Query with EXPLAIN Cost for Open Order Request
Figure 64:	Sample Output
Figure 65:	Optimizer Chose Join Index to Cover Query
Figure 66:	Query with EXPLAIN Cost for Valid Customers with Open Orders
Figure 67:	Valid Customer Answer Set
Figure 68:	Optimizer Chose Join Index for Partial Covering
Figure 69:	Create a Partitioned Join Index
Figure 70:	Using a Join Index with Partition Elimination
Figure 71:	Columns for Multi-Table Join Index
Figure 72:	Eliminate Redistribution and Duplication

Figure 73:	Join Index is Automatically Updated
Figure 74:	Order Join Index
Figure 75:	Compressed Single Table Join Index Syntax
Figure 76:	Non-Compressed Single Table Join Index Syntax
Figure 77:	List Valid Customers with Open Orders
Figure 78:	Query Cost Results
Figure 79:	Vertical Partitioning with a Single Table Join Index
Figure 80:	Store Table Columns
Figure 81:	Store Table and STJI Data Distribution
Figure 82:	Find items sold in a set of stores for a given set of dates
Figure 83:	Create a Join Index that Looks Like the Base Table
Figure 84:	Query that will cause a Full Table Scan of the Join Index
Figure 85:	Syntax for a Value Ordered Index
Figure 86:	Table Data Rows are Hashed Query Does Full Table Scan
Figure 87:	Value Ordered NUSI/STJI – Query scans only specified portion of table based on specified value
Figure 88:	Create a Sparse Join Index
Figure 89:	Create an Aggregated Sparse Join Index
Figure 90:	Table Row Size
Figure 91:	Index Size
Figure 92:	Sparse Index Size
Figure 93:	Sparse Join Index Syntax
Figure 94:	Sparse Single Table Join Index, a Form of Horizontal Partitioning
Figure 95:	Skewed Data Distribution

Figure 96:	12-Month History of all Flyers and all Flights
Figure 97:	Today's Flyers and Today's Flights
Figure 98:	Daily Partition with 100M Rows
Figure 99:	100M Row Subset of the Latest Data
Figure 100:	Create Sparse Join Index for the Year 2011
Figure 101:	Query for 2011 Open Orders
Figure 102:	Query for 2010 Orders
Figure 103:	OrderPPI Table Partitioned By Month
Figure 104:	Sparse Join Index on OrderPPI Table for Q1 2011
Figure 105:	EXPLAIN Output Showing Partition Elimination
Figure 106:	Partitioned Sparse Join Index Syntax
Figure 107:	Partitioned Sparse Join Index with a NUSI Syntax
Figure 108:	Create Customer Table Syntax
Figure 109:	Create Orders Table Syntax
Figure 110:	Create Compressed Global Join Index Syntax
Figure 111:	All AMP Operation Avoids a Full Table Scan
Figure 112:	Global Join Index ("Hashed NUSI") uses a Group AMP Operation
Figure 113:	List the orders for CustId 1500
Figure 114:	EXPLAIN showing use of a Global Join Index as a Hashed NUSI (Partial Listing)
Figure 115:	Compressed Global Join Index
Figure 116:	Non-Compressed Global Join Index
Figure 117:	Determine Space Usage for the Global Join Indexes
Figure 118:	Space Usage Results
Figure 119:	Create Join Index with CURRENT_DATE Syntax
Figure 120:	Join Index with Seven Days of Data

Figure 121:	ALTER Join Index to CURRENT
Figure 122:	Create Hash Index Syntax
Figure 123:	Hash Index ORDER BY Options
Figure 124:	Create NoPI Table Syntax
Figure 125:	Create CustomerTemp NoPI Table
Figure 126:	Create OrderTemp NoPI Table
Figure 127:	Row ID for a NoPI Table
Figure 128:	Multiple NoPI Tables at the AMP Level
Figure 129:	SQL Commands and NoPI Tables
Figure 130:	INDEX LastAccess and AccessCount
Figure 131:	INDEX LastAccess and AccessCount Output

Acronyms / Terms

AJI	Aggregate Join Index
AMP	Access Module Processor
JI	Join Index
MPP	Massively Paralleling Processing
MLPPI	Multi-Level Partitioned Primary Index
MTJI	Multi-Table Join Index
NPPI	Non-Partitioned Primary Index
NUPI	Non-Unique Primary Index
NUSI	Non-Unique Secondary Index
Optimizer	Teradata query optimizer produces the optimal access plan
PDE	Parallel Database Extensions
PE	Parsing Engine
PI	Primary Indexes
PPI	Partitioned Primary Index
SI	Secondary Indexes
STJI	Single Table Join Index
SMP	Symmetric Multi-Processor
Teradata BYNET®	Hardware interprocessor network to link nodes on an MPP system. Note: Single-node SMP systems use a software-configured virtual Teradata BYNET® driver to implement Teradata BYNET® services.
UPI	Unique Primary Index
USI	Unique Secondary Indexes

Table of Contents

Chapter One
TERADATA NODE – A Review 1

Chapter Two
PRIMARY INDEXES 5
    Primary Keys vs. Primary Indexes 5
    Primary Index Criteria 5
    Primary Index Choice Considerations 6
    Primary Index Attributes 8
    How Primary Indexes Work 9

Chapter Three
PARTITIONED PRIMARY INDEXES 11
    How Partitioned Primary Indexes Work 11
    Maximum Partitions 13
    Unique Primary Index Requirements 13
    PPI Basics 13
    Range Queries 14
    Advantages of PPIs 16
    Potential Disadvantages of PPIs 17
    Partition Primary Index Examples 20
    Dates Outside Specified Range 20
    Partitioning Access Options 20
    Partitioning Enhancements 25
    Time-Based PPI Capability 26

Chapter Four
MULTI-LEVEL PARTITIONED PRIMARY INDEX CONCEPTS 29
    Multi-Level Partitioned Primary Index Concepts 29
    Multi-Level Partitioned Primary Index Syntax 31
    Example: Compare District 25 revenue for
    Week 6 vs. same period last year 32

  Multi-Level Partitioned Primary Index with
  AMP-Level Row Grouping 35

  MLPP Insurance Company Example 36

  Calculating the Multi-Level Partition # 39

  CHARACTER Partitioned Primary Index 40

  Changing Multi-Level Partition Definitions 44

  Multi-Level Partitioned Primary Index Rules Summary 45

Chapter Five

SECONDARY INDEXES 47

  Unique Secondary Indexes 48

  Create USI Syntax 48

  Unique Secondary Index (USI) Access 49

  Non-Unique Secondary Indexes 51

  CREATE Non-Unique Secondary Index Syntax 51

  Non-Unique Secondary Index (NUSI) Access 52

  Additional NUSI Criteria 54

  Rows Per Block 54

  Data Distribution 55

  Covered Indexes 55

  NUSI on PI Columns of PPI Tables 59

  Place a USI on NUPI 59

  Place a NUSI on NUPI 62

  Secondary Index Review 64

Chapter Six

JOIN INDEXES 67

  SINGLE TABLE VS. MULTI-TABLE JOIN INDEX OVERVIEW 68

  OPTIONS FOR JOIN INDEXES 69

  Sparse Join Index 69

  Global Join Index 69

  Join Index Considerations 69

  Join Index Example – Customer and Orders Tables 70

Compressed Join Index 72

Non-Compressed Multi-Table Join Index 74

Compressed and Non-Compressed
Join Index Storage Comparison.. 75

Will a Compressed Join Index Help Access? 79

Teradata Database Join Index Optimization.. 84

Multi-Table Join Index Usage 84

Creating a Single Table Join Index 87

Index Covering with a Single Table Join Index (STJI) 90

Single Table Join Indexes and Non-Unique Secondary Indexes 90

Base Table and Single Table Join Index with
Different Primary Indexes 92

Same Primary Index for Single Table Join Index –
Acts Like a NUSI 94

Global Join Index – Using the LIKE Clause on a STJI 95

Build a STJI with a Column for LIKE Plus RowID of Base Table 96

Value Ordered Indexes for Range Searches 97

SPARSE Join Indexes 99

Sparse Single Table Join Index –
A Form of Horizontal Partitioning 103

Sparse Single Table Join Index – Three More Examples..................... 105

Value Ordered Sparse Single Table Join Index 105

Value Ordered Sparse STJI on PPI Table 107

Creating a Compressed Sparse Multi-Table Join Index 108

Creating a Sparse Join Index on a Partitioned Table 110

Creating a Sparse Join Index with Partitioning 112

Global Join Index on Multiple Tables 113

Global Index as a "Hashed NUSI"..115

Repeating Row IDs in Global Join Index 119

Temporal Support – Move Date in Join Index 121

Join Index Type Review 123

Chapter Seven

HASH INDEXES 125
    Hash Index Example 127

Chapter Eight

NO PRIMARY INDEX TABLES 129
    NoPI Table Load 129
    CREATE NoPI TABLE with New Syntax 130
    The Row ID for a NoPI Table 131
    AMP-Level View of Multiple NoPI Tables 132
    Populating a NoPI Table from Existing Tables 133
    Populating a PI Table from an Existing NoPI Table 133
    SQL Commands and NoPI Tables 134

Chapter Nine

INDEX REVIEW 137
FAQs 139
Index 145

Chapter One
TERADATA NODE – A Review

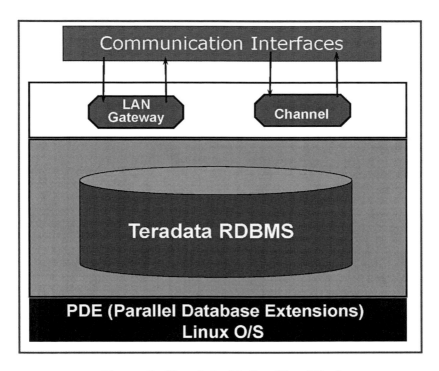

Figure 1: Teradata Node - Simplified

As a very basic review, let's begin with a Teradata node. The operating system currently being deployed is Linux®. Older models may still be running on 32-bit UNIX® or Windows, but Linux is 64-bit and Teradata can take advantage of the expanded memory offered by Linux.

There is a layer of software called Parallel Database Extensions (PDE) that insulates the Teradata Database from the operating system. Teradata Database asks PDE for needed resources and PDE manages the request to the operating system. Linux is unaware that Teradata Database is running on it.

The Teradata Database software manages Teradata's massively parallel data warehouse environment, including the virtual processors and the Teradata BYNET® message passing layer. The Teradata BYNET® is what makes Teradata run in parallel and it possesses high-speed logic that provides bi-directional broadcast, multicast, and point-to-point communication between the virtual processors, plus the Teradata BYNET® provides merge functionality.

The other main piece of software in a node is the communications interface that allows the Teradata Database to communicate with "the outside world."

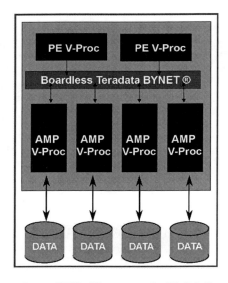

Figure 2: Teradata SMP (Symmetric Multi-Processor) Node

A single node is called an SMP server, a symmetric multi-processor. The two types of virtual processors it contains are called Parsing Engines (PEs) and Access Module Processors (AMPs). User requests come into the node through the PE. The SQL query is parsed and the code is optimized. The resulting processing steps are sent via the Teradata BYNET® to the AMP or AMPs that retrieve and process the requested data.

The Teradata Database is a parallel environment. That means the data is distributed across all AMPs, Teradata Database's units of parallelism.

The data from a table is spread evenly and randomly across all the AMPs on the system, not just stored on a single AMP. Each AMP has its own virtual data storage. Physically there are many drives, however, the AMP only "sees" a list of disk cylinders and does not make a distinction as to how many physical drives are attached to it.

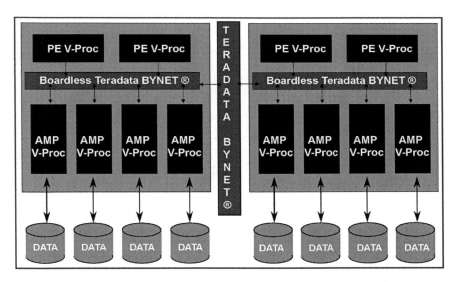

Figure 3: Teradata MPP (Massively Parallel Processing) Nodes

A single node system requires no physical TERADATA BYNET®, it has a virtual TERADATA BYNET® instead. A system of two or more nodes requires a physical Teradata BYNET® to connect the nodes. For simplicity, the example in Figure 3 shows four AMPs in the node. A real system generally contains somewhere in the neighborhood of 40 AMPs per node. Nodes support up to 128 AMPs, however, systems are not configured to the maximum number of AMPs for failover capability.

When expanding a Teradata Database, the Teradata BYNET® is increased to support additional nodes. Data is moved from existing AMPs to new AMPs. Data is never moved from an existing AMP to another existing AMP. This minimizes time, cost, and processing.

When we need to access data, we want to do it in the most efficient manner. Indexes are the answer.

This book covers the following Indexes:

- Primary Indexes
- Partitioned Primary Indexes
- Multi-Level PPI both numeric and character
- Unique and Non-unique Secondary Indexes.

We'll follow that with other Indexing scenarios:

- Single Table Join Index
- Value Ordered Index
- Sparse Join Index
- Global Join Index
- Value Ordered Sparse STJI
- Multi-Table Join Index
- Hash Indexes
- No Primary Index Tables

The last section includes Frequently Asked Questions and a sample query to use when determining the last time indexes were accessed.

In all cases, the intent is to explain what the indexes are, how they work, and how they can be used to access data most efficiently.

Chapter Two
PRIMARY INDEXES

As we strive to get the best possible response times for our queries, we often find ourselves searching for more efficient methods to use our data warehouses. Teradata Database indexes provide numerous physical access paths to the data, with many alternatives and variations, which can significantly improve performance. Primary Indexes provide the best access to the Teradata Database.

In this section we will explore how Unique and Non-Unique Primary Indexes are used to both store and access data.

Primary Keys vs. Primary Indexes

Primary Indexes can be unique or non-unique so they may or may not be the same as the Primary Key from the logical data model. The Primary Key is the unique identifier of a row in a logical model table, whereas the Primary Index determines the physical location of where the row is stored in the system.

Primary Index Criteria

The choice of a table's Primary Index (PI) is of the utmost importance as it can make or break the design of the data warehouse.

What are the criteria for choosing a Primary Index? The three main criteria to consider when choosing a table's Primary Index are:

1. Access
2. Distribution
3. Volatility

There have been many discussions over the years about which is more important; "access" or "distribution". In fact, most people would say data distribution is the number one criteria, and for most Relational Database Management Systems (RDBMs), they would be correct. The answer for a Teradata Database is to consider both access and data distribution.

The choice of Primary Index truly depends on the entire context of the data warehouse environment because the Primary Index determines the AMP where the row will be stored and, at the same time, represents the most efficient access path to the data. In fact, access is actually more important in a Teradata Database environment, but we must still pay close attention to how the data distributes across the AMPs.

The two ways to look at access are value access and join access. Value access is when a column is listed in the WHERE clause, such as color = 'red'. Value access criteria can also be met with Secondary Indexes which are discussed in a later chapter.

Join access is when a column is used to join two tables such as Claim.StateId = State.StateId. Join access and Primary Indexes are the key contributors to making joins run efficiently. Join access can also be improved by Non-Unique Secondary Index (NUSI) access.

Primary Index Choice Considerations

At this point, we know that the Primary Index determines which AMP a row is stored on. Let's see how Primary Indexes provide the most efficient access to those rows.

If a value for the Primary Index (a value the optimizer can hash) is not supplied in the WHERE clause to access the data, then Full Table Scans (or other indexes) will be used. Potentially critical workloads that require short response times could take much longer than service level agreements, as the system reads every row in a table.

If a table is used to join to another table that has a different Primary Index, one or both of those tables will have to be redistributed across the AMPs in order to be able to perform the join. Joining rows on the Teradata Database is much like putting a jigsaw puzzle together. Puzzle pieces are moved around until a match is found, then the pieces can to be physically joined together to form a larger piece. It is the same with data rows. We may have to move rows around the system to different AMPs to find a match for the column we are joining on. When we consider the join criteria for our tables, the choice of Primary Index can avoid data movement before the rows are physically joined.

It is true that data should be distributed as evenly as possible across the AMPs so they have an even workload, but consideration must first be given to how the data will be accessed. Therefore, it is more important to consider how the table is accessed than how the data is distributed, and we have to compare the two simultaneously.

If there is a requirement to maintain the Primary Key uniqueness from the logical data model, then we may choose to use a Unique Primary Index (UPI). The Teradata Database software uses an algorithm based on actual user data in the indexed column(s) and table rows are evenly and randomly distributed across all the AMPs. We will see later in the book that a Unique Secondary Index (USI) can also be used to maintain data uniqueness.

The Teradata Database is also quite efficient when a table has a Non-Unique Primary Index (NUPI), especially if the values are nearly unique. Again, placement of rows on the AMPs is important when it comes to accessing the data and for certain circumstances, NUPIs can be even more efficient than UPIs.

Lastly, we must consider index column data volatility. If a data value changes frequently, then it may not be a good choice for the Primary Index. The original row will have to be deleted and a new row inserted on a different AMP when the value of the Primary Index is changed. We call this an "unreasonable" update.

Ultimately, the optimizer will make the choice of which index to use based on the number of nodes in the system, the number and type of CPU's per node, the number AMPs, the disk array configuration, amount of memory, columns with indexes, number of rows in the table, number of rows per data block, number of values per column, and number of rows per value.

Primary Index Attributes

The PI is meant to be the primary access path to the data, but usage determines if that is so. When used for access, the Primary Index is the most efficient possible access. A table has one and only one PI for its lifetime. The Primary Index cannot be changed without creating a new table. Of course, the actual data in the PI column(s) can be changed by updating the data, but then again, volatile data is not a good choice for a PI as each time the data changes, so does the row's physical location.

- The primary physical access path
- The mechanism used to assign a row to an AMP
- A table must have one and only one Primary Index
- Primary Index cannot be changed without recreating the table
- UPIs result in even distribution of the rows of the table across all AMPs.
- UPIs ensure no duplicate rows
- PI accesses are always one-AMP operations
- NUPIs will result in even distribution of the table rows proportional to the degree of uniqueness of the index and the number of AMPs
- Primary Indexes may or may not be the same as Primary Keys

Figure 4: Primary Index Attributes

A Unique Primary Index (UPI) gives very even distribution of data across the AMPs. The system ensures there are no duplicate rows when a table has a Unique Primary Index. UPI access is great for tactical queries that require very quick response times.

When accessing a table using a single Primary Index value, the system only accesses one AMP to retrieve the data.

If the Primary Index is Non-Unique (NUPI), the data may somewhat unevenly distribute across the AMPs. The unevenness of the distribution is determined by the degree of uniqueness of the data and the number of AMPs on the system. A NUPI with more unique values distributes the data more evenly.

How Primary Indexes Work

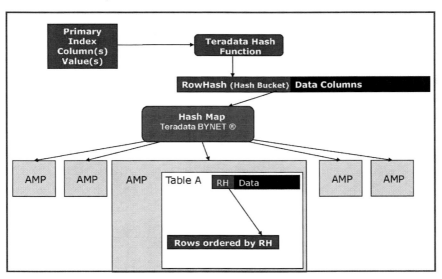

Figure 5: Primary Index RowHash Determines AMP Location for Storage

This is how the Primary Index works: when the table is defined, a column or group of columns is chosen as the Primary Index and that index may be unique or non-unique. The indexed column(s) is part of the user data row and is not replicated. In the PE, when a row is being formatted, the value of the Primary Index column(s) is run through the Teradata Hash Function where the algorithm produces a 32-bit row hash based on the data values and data types. The RowHash is added to the beginning of the data row. Depending upon the release of Teradata Database software and the hashing algorithm employed, either the first 16 or 20 bits are used for bucketing.

The Teradata BYNET® owns the hash map and from the RowHash, Teradata BYNET® software determines which AMP owns the bucket number. Once a row is sent to its designated AMP, the Teradata File System determines the physical placement of the row on the AMP's corresponding storage.

Inside a data block, the rows are stored in row hash sequence and an additional 32-bits counter is added to the row hash to form the RowID. A RowID uniquely identifies a row within a table. For Primary Index access, the row hash is used to locate the row. Later we will see how the RowID is used by other indexes to locate that same row.

Chapter Three
PARTITIONED PRIMARY INDEXES

In addition to Primary Indexes, which in a broad sense can be described as vertical partitioning, Teradata Database also offers the ability to partition the data horizontally through the use of Partitioned Primary Indexes (PPI). Partitioning provides access to the portions of the table we want to use and eliminates access to the partitions that do not satisfy our query, reducing physical I/O and saving system resources.

Functionally, we still use the Primary Index. We still run the value through the hashing algorithm and we still assign the row to its AMP as we do for any Primary Index.

Within the AMP, partitioning additionally allows various rows to be grouped together based on values and ranges of values of the "partitioning column(s)." Partitioned Primary Indexes can, therefore, optimize the physical database design for range constrained queries.

How Partitioned Primary Indexes Work

The partition is a grouping of like rows. For example, we have a Sales Table with a Primary Index on CustId and we partition on the month column. All the rows for the same month are grouped together on their respective AMPs, and striped across all the disks, because each AMP has all the partitions. The rows are assigned to AMPs based on the row hash of the Primary Index value, and then grouped by partition (RowKey). Within partitions, rows are ordered by the row hash of the Primary Index.

Teradata® Database Index Essentials

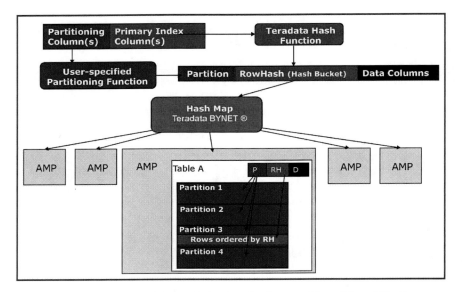

Figure 6: Partitioned Primary Indexes: How They Work

A feature of partitioning is the ability to eliminate partitions from being retrieved. This makes access to the data much faster, because data blocks are not read for the partitions that can be eliminated for a query. For example: If a Sales Table has data for five years, and it is partitioned by month, it will have 60 partitions. If we only access one month, then 59 of the 60 partitions are eliminated. Partition elimination creates significant performance improvement.

PPI tables are easy to manage. Partitions are defined when the table is created. When data is loaded, the Teradata Database maintains the partitions automatically. High-volume batch insert times are reduced by as much as 90% when using Partitioned Primary Indexes. Deleting large numbers of rows from PPI tables is nearly instantaneous.

Maximum Partitions

When creating a table, there is a maximum of 65,535 partitions for a table. Multiple columns can be named in the partitioning expression. In Teradata Release 14, the number of PPI partitions is (2^{63} - 1), 9.2 quintillion.

9,223,372,036,854,775,807

The increased partition limit will expand the size of the RowKey.

Unique Primary Index Requirements

If a UPI is required on a partitioned table for better performance, then the partitioning columns must be defined as part of the Primary Index, otherwise, PPI tables are defined with Non-Unique Primary Indexes.

PPI Basics

To review the basics, the rows are still hashed by the Primary Index column. Once a row gets to its AMP, it is ordered by partition and by RowHash within the partition.

PPIs are allowed on base tables, Global Temporary tables, Volatile Temporary tables, and Non-Compressed Join Indexes.

Not all tables are suited for having partitioning. For example, if the partitioning column is not qualified in the WHERE clause of a SQL statement, all the partitions will have to be read, causing response time degradation.

Another performance issue could occur when joining on PI columns from a non-PPI table to a PPI table. This can result in looking at every partition of the PPI table to compare the PI column, resulting in a more costly operation.

By adding partitioning, we can often drop unneeded Secondary Indexes or Value Ordered Join Indexes (to be explained in a future chapter).

Strategic queries look at all the data in the table. Tactical queries look at only the subset of data they need to satisfy a query, therefore tactical queries can benefit from partitioning.

Range Queries

Consider two tables, one with a Primary Index and the other with a Partitioned Primary Index. Executing the query in Figure 7 against these two tables, where we access September dates, will generate two different access paths.

```
SELECT   MyColumn
FROM     MyTable
WHERE    OrderDate
BETWEEN DATE '2011-09-01' and DATE '2011-09-30';
```

Figure 7: PI Range Query

For the Non-PPI table, Figure 8 shows how every row is read on each of the AMPs, searching for all the September dates.

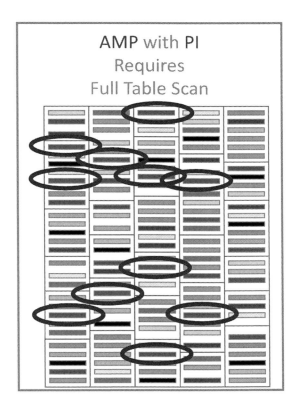

Figure 8: Full Table Scan on AMP with PI

For the PPI table in Figure 9, all the September rows are grouped together. Each AMP only accesses the September partition, and eliminates the remaining partitions. None of the non-September rows are not accessed. This is more significant as more years of data are added to the table.

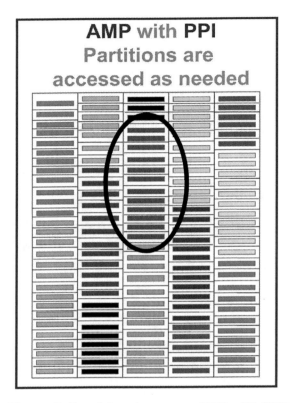

Figure 9: Partition Access on AMP with PPI

Advantages of PPIs

Remember, not every table is a candidate to have a Partitioned Primary Index. The best practice of testing various alternatives with the EXPLAIN Facility is recommended to get the maximum benefit from PPIs. There are some trade-offs to consider.

Partition elimination significantly reduces I/O and system usage. While access is still an all AMP operation, partitioning reduces the amount of data that needs to be scanned on each AMP.

Partitions allow fine granularity on the separation of data. This allows us to focus on the data we want to see without looking at the entire table. Creating smaller partitions, i.e. day rather than month or month rather than year, allow us to scan less data.

Partitioned Primary Indexes may eliminate the need for some secondary indexes that would have been defined on the partition. In this case, the overhead for the secondary index is eliminated.

Partition deletion is nearly instantaneous, because it is block level, rather than the row level. If there are other indexes on the table, the partition deletion is still very fast, but the deletion of the index entries can take a significant amount of time.

PPIs can result in dramatic improvements in query response times, in high-volume data loads, and maintenance operations. It is imperative that database administrators understand the trade-offs when considering using PPIs on tables.

Potential Disadvantages of PPIs

Each row of a PPI table has a two byte overhead to accommodate the RowKey information about the partition.

The Primary Index cannot be defined as unique, unless the partitioning column(s) is defined as part of the Primary Index.

To take advantage of a multi-column Primary Index, all the columns must be specified in the WHERE clause with a set of values that can be hashed by the optimizer. The Teradata Database does not do partial index searches.

Access can be degraded if the partitioning column is not specified in the WHERE clause of the query. If we join a PPI table to a non-partitioned table with the same PI, those joins may also be degraded.

In the past, the Teradata query optimizer had to see the partitioning column as a constant, to determine which partitions to exclude.

With techniques such as Constant Folding, Satisfiability (SAT), and Transitive Closure (TC) added in Teradata Releases 12 and 13, the optimizer does constant arithmetic on expressions and does partition elimination. A common example is DATE – x, where x is a constant number of days. The optimizer can also handle some less obvious cases, where it moves constants across a comparison operator to get them together and compute a constant value to use for the partition. An example is: WHERE Date_Column + interval '3' month <= '2011-03-31' becomes WHERE Date_Column <= '2010-12-31'. There is also a caution when doing range partition deletes. If data falls outside the defined partition ranges, that data will be moved to the NO RANGE partition and the move will not be fast because the data will be sent to a "remote" location on the disk drive.

Partitioned columns may not be compressed.

```
Partition the 1.2 Million row Claim table by "ClaimDate".
CREATE TABLE Claim
    ( c_ClaimId      INTEGER  NOT NULL
    ,c_CustId        INTEGER  NOT NULL
    ,c_ClaimDate              DATE        NOT NULL
    ... )
PRIMARY INDEX (c_ClaimId)
    PARTITION BY RANGE_N (c_ClaimDate BETWEEN
    DATE '2002-01-01' AND DATE '2011-12-31' EACH INTERVAL '1' MONTH );
```

Figure 10: Partitioning with Range _N

The table in Figure 10 has a ClaimId, CustId, ClaimDate, and various other columns. The ClaimId values are unique. Because the table is partitioned on ClaimDate, we cannot define the ClaimId column as unique. ClaimDate has to be part of the Primary Index definition if we want a Unique Primary Index on this table.

Partitioning expressions may use either a RANGE_N or CASE_N function. RANGE_N defines a range of boundaries. It is the most common type of Teradata Database partitioning expression. A sample definition is: RANGE_N SaleDate BETWEEN DATE '2010-01-01' AND DATE '2011-12-31' EACH INTERVAL '1' MONTH.

CASE_N is used to evaluate a list of conditions. It will return the first condition that evaluates to true, unless a prior condition in the list evaluates to UNKNOWN. A sample definition is CASE_N (TotalSales < 1000, TotalSales < 10000, NO CASE, UNKNOWN).

The Claim Table has the Primary Index on ClaimId and the ClaimDate is partitioned by RANGE_N. The range of dates is from 2002 through 2011. Sectioning each partition into one month intervals, gives ten years of information in 120 monthly partitions.

The table is empty when it is created. To insert large amounts of data, tables are populated using a Teradata utility such as FastLoad or MultiLoad. For this example (Figure 11), we use ad-hoc inserts to add two data rows.

```
INSERT INTO Claim VALUES (100039,1009, '2002-01-13', ...);
→ placed in partition #1

INSERT INTO Claim VALUES (260221,1020, '2011-01-07', ...);
→ placed in partition #109
```

Figure 11: Partition Inserts

We insert values for the ClaimId, CustomerId, and a ClaimDate into the Claim Table. The first date, January 13, 2002, corresponds to the first month of the 120 partitions. The Parsing Engine determines that this date gets stored in partition #1, and communicates this information to the AMP.

If we insert the row for January 7, 2011, it is inserted into partition #109. As data is inserted, each row is evaluated and stored in the appropriate partition.

It then follows that a row for December 2011 will be stored in partition #120, the last partition we have defined.

A few inserts will not work with this table. For a claim with a date of 1999, there is no defined partition. That is, we have no partition for values outside the range of defined partitions, and it will return a partitioning violation message as seen in Figure 12.

> 5728: Partitioning violation for table ATLCLH.Claim.

Figure 12: Partitioning Error Message

Dates Outside Specified Range

For the case of prior or future years, Teradata foresaw the need to handle dates outside a specified range.

The No_Range and Unknown options support dates that are either NULL or out of range.

Partitioning Access Options

Let's see what happens when we take advantage of partitioning. Teradata Database does not set aside separate blocks or separate areas for different partitions. Partitions are ordered within an AMP and within a partition, rows are ordered by the row hash.

The following examples use a Claim Table with approximately 1.2 million rows on a small Teradata system. The query in Figure13 accesses the Claim_PPI Table, where the ClaimDate is between January 1 and January 31, 2011.

If the table is not partitioned, we would have to do a Full Table Scan. With a Partitioned Primary Index (Figure 13) we do an all-AMP operation, but only scan a single partition.

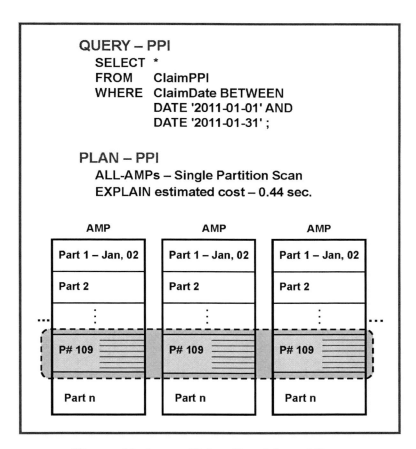

Figure 13: Access Using Partitioned Data

Partition #109 will be returned and 119 partitions will be eliminated. The actual EXPLAIN plan shows the cost as 0.44 seconds, or less than one second to return all of the claims for the month of January 2011. Just a note on the EXPLAIN "time" estimate. The optimizer is cost based. The numbers generated by the optimizer are cost estimates that can be compared to other cost estimates. The EXPLAIN labels these numbers as "time", however, they are not intended to represent execution times, they represent a cost estimation. Cost estimates from one query can be compared to cost estimate for another query.

Now let's compare the results of accessing a PPI table to a table (Figure 14) that is not partitioned.

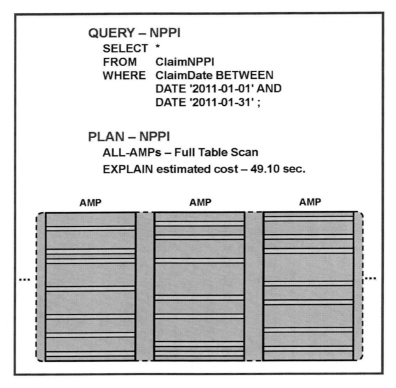

Figure 14: Access Using Non-Partitioned Data

This still results in an all AMP operation, but now it requires doing a Full Table Scan. The EXPLAIN shows a cost of almost 50 seconds compared to the previous query cost of .44 seconds. Clearly, partitioning helps this kind of query.

Partitioning clearly helps some queries, but it may hurt other types of queries.

Let's select ClaimId 260221 from the Claim_PPI Table in Figure 15. Using the ClaimId, the Non-Unique Primary Index of the table, makes this is a one-AMP operation. We go to the AMP, but the RowHash for 260221 can appear in any one of the partitions. This necessitates doing a "partition lookup". We do not have to scan all the data on the AMP. Using the specific RowHash related to the ClaimId value of 260221, we read the memory-resident master index, then read a cylinder index, and finally the data block where the first partition is stored.

We must access each of the partitions on the AMP to locate the RowHash we are looking for until we find it. This example has 120 partitions, so we might have to do 120 physical I/Os to find the desired ClaimId. The EXPLAIN cost shows that this query has a relatively low cost of 0.09 seconds.

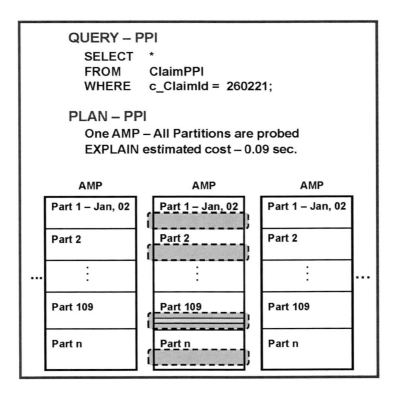

Figure 15: Access Using a PPI table via the Primary Index with No Partition Specified

Now, contrast that example with the following non-partitioned table (Figure 16). To satisfy a Unique Primary Index query, we access one AMP. A non-partitioned table is considered to have one large partition, called partition #0. Using the file system, we can read one data block, and access one data row. The EXPLAIN cost is 0.00 seconds, although the query execution takes a tiny bit of time.

It is a slow process when Primary Index access requires looking at all the partitions. In another section we shall demonstrate a technique that helps this type of Primary Index access run faster. We will discuss how adding a Unique Secondary Index (USI) or a Non-Unique Secondary Index (NUSI) on the Primary Index of PPI tables improves performance.

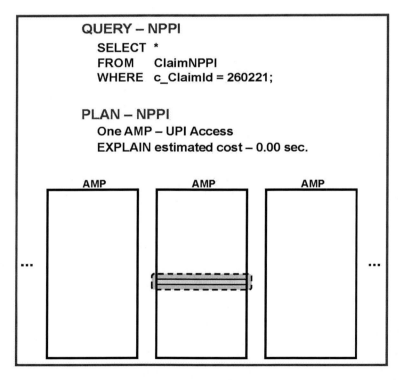

Figure 16: Accessing a NPPI table via the Primary Index

Partitioning Enhancements

Teradata Database has a number of enhancements that have been made to partitioning over the last decade (Figure 17). Partitioning is a very widely used feature within the Teradata Database, however, only a select number of tables are partitioned. Partitioning on key tables benefits us significantly. The most common type of partitioning is date type partitioning with RANGE_N.

> **Teradata V2R6.0**
> - Single-AMP NUSI access when NUSI on same columns as NUPI
> - Partition elimination on RowIDs referenced by NUSI
>
> **Teradata V2R6.2**
> - PPI for non-compressed join indexes
>
> **Teradata 12.0**
> - Multi-level partitioning
>
> **Teradata 13.10**
> - Tables and non-compressed join indexes can now include partitioning on a character column.
> - A PPI table now allows a test value (e.g., RANGE_N) to have a TIMESTAMP(n) data type.

Figure 17: Various PPI Enhancements

Teradata Database provides the ability to create a Secondary Index on a Non-Unique Primary Index, and partition elimination on RowIds referenced by NUSIs.

Non-Compressed Join Indexes may be partitioned.

In Teradata Release 12, multi-level partitioning was added. Teradata Release 13.10 offers character and graphic partitioning, plus a TIMESTAMP data type.

Time-Based PPI Capability

When Teradata added the temporal feature, PPIs were enhanced to allow execution of an Alter Table command. Previously populated PPI tables that are either defined with CURRENT_DATE or CURRENT TIMESTAMP (Figure 18) can be altered (Figure 19) to redefine the table partitions.

```
CREATE TABLE TimeBasedPPI
        (ProductId        INTEGER,
        ProductVal        INTEGER,
        PartitionDateCol  INTEGER)
PRIMARY INDEX (ProductId)
PARTITION BY CASE_N
( PartitionDateCol >= CURRENT_DATE - INTERVAL '2' DAY,
  PartitionDateCol >= CURRENT_DATE – INTERVAL '90' DAY,
  PartitionDateCol >= CURRENT_DATE – INTERVAL '180' DAY,
  PartitionDateCol >= CURRENT_DATE – INTERVAL '270' DAY,
  NO CASE);
```

Figure 18: Create Time-Based PPI Table

Partitions are set based on the current date. The first partition has two days of data; the next is an offset of 90 days from the current date, followed by partitions with intervals of 180 and 270 days. As new data is added to the table, the first partition is populated. Over time, that partition grows and contains more than two days of data. Since queries tend to be against recent data, as time passes, the partitions need to be adjusted to reflect more current information (Figure 20) and limit or reduce the size of recent partitions.

By altering the table, data in the referenced partitions is moved according to the defined date intervals. As a result, the first partition will again contain two days of data and queries against recent dates have to scan less data.

ALTER TABLE TimeBasedPPI TO CURRENT;

Figure 19: Alter Time-Based PPI Table

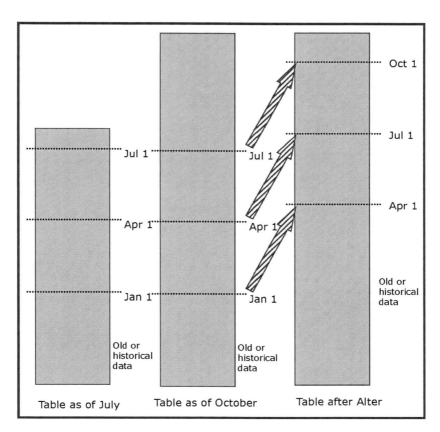

Figure 20: Alter Effect on Time-Based PPI Table

Altering the table does not repartition it, but the data is moved to realign with the CURRENT_DATE definitions of the table. Although it is fairly unusual to have different partition granularity and the ALTER may be quite costly depending upon the number of rows involved, this example illustrates how the feature works in the Teradata Database.

Chapter Four
MULTI-LEVEL PARTITIONED PRIMARY INDEX CONCEPTS

ML-PPI on table(s) gives the Teradata optimizer a greater degree of partition elimination at a more granular level, creating a greater level of query performance.

Teradata Database also offers Multi-Level Partitioned Primary Indexes which are an extension to the PPI capability. Multi-Level Partitioned Primary Index (ML-PPI) defines up to 15 levels of partitions. A ML-PPI can be specified for a table or Non-Compressed Join Index. This optimizes the degree of partition elimination at an even more granular level, resulting in much better query performance.

The optimizer determines whether or not to use the index and partitioning. It uses an index as part of the planned query execution automatically. The question is, can the optimizer use partitioning and partition elimination, to further enhance query performance?

Multi-Level Partitioned Primary Index Concepts

Multi-level partitioning allows each partition to have one sub-level partition up to 15 levels deep. ML-PPI works the same as the single level partition where the RANGE_N or CASE_N expressions are also available. ML-PPI allows us to define multiple WHERE clause predicates that result in additional partition elimination.

Internally, the range partitions are combined into a single partitioning expression and that defines how the data is partitioned on the AMP. If we use a PARTITION BY expression, and it is specified for the Primary Index, that index is called a Partitioned Primary Index.

A PPI specifies a single partitioning expression. For PPI tables, the rows continue to be distributed across the AMPs based on the Primary Index, but within the AMP the rows are ordered by the partition # and then within the partition, by the RowHash of the Primary Index.

ML-PPIs specify more than one partitioning expression. Rows continue to be distributed across the AMPs based on the Primary Index. The rest of the process is the same as for PPIs, only now there are multiple partitions.

The Teradata Database can take a main level partition and sub partition it down to 14 sub levels for a total of 15 total partition levels. At this point in time, it is doubtful that we will use that many levels, because through Teradata Release 13.10, there is a maximum of 65,535 partitions. This limit is expanded to 9.2 quintillion partitions in Teradata Release 14. Referring to the Claim Table in Figure 10, which had 120 partitions for 10 years of 12 months of data; let's say that table also includes a location column with 100 locations. If we also partition on the location column, then we have 120 months of data for each of the 100 locations. We then multiply 120 * 100 giving us a total of 12,000 partitions. That is easily accommodated within the limit of 65,535 partitions.

If the interval is for one day instead of one month, then ten years of information at 365 days per year, plus two leap years, requires 3,652 daily partitions. Then, when we multiply the 100 locations by the 3,652 days, we need 365,200 and that exceeds our limit of 65,535 partitions by 299,664 and we see that this level of partitioning does not work.

The number of partitions at level 1 multiplied by number of partitions at level 2 multiplied by ... the number of partitions at level 15 is equal to the total number of partitions allowed on a table, not to exceed 65,535 partitions.

Multi-Level Partitioned Primary Index Syntax

This is the syntax that is needed to support Multi-Level Partitioning.

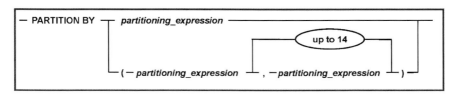

Figure 21: Multi-Level Partitioned Primary Index Syntax

Syntactically, for only one level of partitioning, no parenthesis are needed for the partitioning expression. For multiple levels of partitioning, the entire partitioning expression must be placed within parenthesis. The partitioning expressions can be in any order. If a partitioning expression will need to have its number of partitions expanded, it must be defined as the top level.

Example: Compare District 25 revenue for Week 6 vs. same period last year

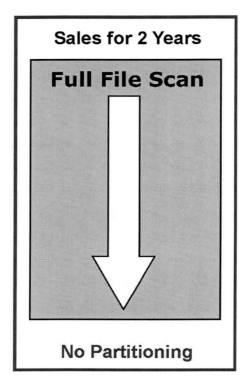

Figure 22: Sales for 2 Years With No Partitioning

Let's examine an example of how to take advantage of multi-level partitioning. For the first example, we have sales information for two years. If there is no partitioning, and we select District 25 revenue for Week 6, versus the same week one year ago for the same District, a Full Table Scan is required to access the data.

Figure 23: Week 6 Sales Only and Single Level Partitioning

In this next example, there are 24 partitions, each representing one month for a two-year period. To access a specific week, we look at the entire month. Since we are looking at only two of the 24 months, we only access 1/12 of the data.

We derive extra benefit by dividing sales into districts. To keep the numbers easy, let's say we have 50 districts. We multiply 24 months times 50 districts, and that gives us a total of 1,200 partitions. Now, select a specific district for a specific week. Look at the month for that district and we only have to access two partitions out of 1,200. This query could potentially run 600 times faster, because we only look at 1/600 of the data. And that, dear reader, is the benefit of multi-level partitioning.

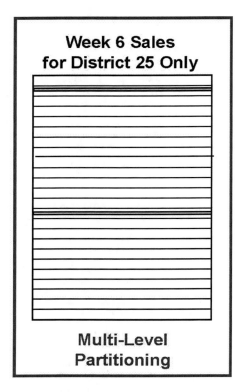

Figure 24: Week 6 Sales for District 25 and Multi-Level Partitioning

Instead of partitioning by month, we can partition by week or by day and instead of partitioning by district, we can partition by store. Keep in mind that we have to multiply the total number of partitions for each level, without exceeding 65,535, until Teradata Release 14.

Multi-Level Partitioned Primary Index with AMP-Level Row Grouping

Now we have an example of a Sales Table with three levels of partitioning.

```
CREATE TABLE Sales
   (StoreId        INTEGER NOT NULL
   ,ProductId      INTEGER NOT NULL
   ,SalesDate      DATE FORMAT 'yyyy-mm-dd' NOT NULL
   ,TotalRevenue   DECIMAL(13,2)
   ,TotalSold      INTEGER
   ,Note           VARCHAR(256))
UNIQUE PRIMARY INDEX (StoreId, ProductId, SalesDate)
   PARTITION BY (
      RANGE_N(SalesDate BETWEEN
         DATE '2003-01-01' AND DATE '2005-12-31' EACH
         INTERVAL '1' YEAR).
      RANGE_N(StoreId BETWEEN 1 AND 300 EACH 100),
      RANGE_N(ProductId BETWEEN 1 AND 400 EACH 100));
```

Figure 25: CREATE Table Statement for Unique Partitioned Primary Index with three levels

The Sales Table has StoreId, ProductId, SalesDate, TotalRevenue, TotalSold, and a Note column. It has a Unique Primary Index, which means all of the partitioning columns must be part of the Primary Index definition.

For the partitions, we define RANGE_N SalesDate for a range of three years at one-year intervals, giving us three partitions. The next level is RANGE_N StoreId for 300 stores at 100 store intervals which creates three partitions. The third level partition is RANGE_N ProductId for 400 products at intervals of 100 which is four partitions. How many partitions do we actually have? The total number of partitions is 36, calculated by multiplying 3 years times 3 stores multiplied by 4 products.

Part. #	L1	L2	L3	Sales					
				StoreId	ProductId	SalesDate	Total Revenue	TotalSold	Note
1	1	1	1	96	10	2003-04-15	4158	42	Good day
2	1	1	2	71	184	2003-07-06	1972	68	Marginal
3	1	1	3	80	241	2003-11-09	3055	47	Slow day
4	1	1	4	82	363	2003-12-24	1261	13	Promotion

Figure 26: Sample Row Header for Sales Table with Three Levels of Partitioning

The row format in the Figure 26 shows there is partition information associated with each row. The row overhead, even for multiple levels of partitioning, is still only two bytes.

MLPP Insurance Company Example

An insurance company often performs analysis for a specific state and within a date range that is a small percentage of the many years of claims history in their data warehouse. The object is to attain partition elimination, using multiple expressions for filtering, based on the WHERE clause predicates, to benefit performance.

If analysis is being performed for Vermont claims, or claims in January 2011, or Vermont claims in January 2011, creating data partitions that allow for the elimination of all but the desired claims, has an extreme performance advantage.

It should be noted that the ML-PPI provides direct access to partitions, regardless of the number of levels specified in the query. This assures partition elimination and enhanced query performance.

Now let's take the Claim Table and the 10 years of information and subdivide this into states.

```
CREATE TABLE Claim
    ( ClaimId      INTEGER  NOT NULL
    ,CustId       INTEGER  NOT NULL
    ,ClaimDate    DATE     NOT NULL
    ,StateId      BYTEINT  NOT NULL,
    ... )
PRIMARY INDEX (ClaimId)
    PARTITION BY (
        /* First level of partitioning */
        RANGE_N (ClaimDate BETWEEN
            DATE '2002-01-01' AND DATE '2011-12-31'
            EACH INTERVAL '1' MONTH ),
        /* Second level of partitioning */
        RANGE_N (StateId BETWEEN 1 AND 75 EACH 1) )
UNIQUE INDEX (ClaimId);
```

Figure 27: Partition Claim Table by ClaimDate and StateID

The whole partitioning expression is contained inside parenthesis. RANGE_N ClaimDate has 120 partitions and is subdivided by StateId. RANGE_N StateId defines 75 states. There are 50 states in the USA, not 75, however, we use a lot of StateIds that do not necessarily represent states. For example, the District of Columbia, the Virgin Islands, and the Armed Forces of America have StateIds. For that reason, we define 75 partitions to accommodate all the additional StateIds.

Now, the question is how many partitions do we have defined? We still have ten years partitioned by one month, which is 120 month (ClaimDate) partitions. Now we subdivide each of those 120 months into 75 StateIds. The Claim Table has a total of 9,000 partitions defined on it.

Teradata Database can take advantage of a multi-level partitioning expression in a number of ways. It can select data for any individual level, or any combination of levels, which need not be contiguous.

```
SELECT    ...
FROM      Claim C, States S
WHERE     C.StateId = S.StateId
AND       S.StateName = 'Vermont'
AND       C.ClaimDate BETWEEN DATE '2011-01-01'
          AND DATE '2011-01-31';
```

Figure 28: Access a Multi-Level Partitioned Primary Index Table

We have StateId in the Claim Table. Our enterprise also has a look up table for State information with columns for the StateName, the StateAbbreviationCode (e.g., VT for Vermont), and a StateId number. The State lookup table joins to the Claim Table, matching on the StateId where the StateName is Vermont. Optionally, we can request a specific range of values. The range we are looking at is claims for January 2011.

Combining both the WHERE clause predicates, searching for Vermont claims in January of 2011, results in scanning less than 0.02% of the data.

Eliminating all but one month out of many years of claims history facilitates scanning less than 1% of the claims history. Eliminating all but the Vermont claims out of the many states facilitates scanning less than 2% of the claims history. Combining both of these predicates for partition elimination facilitates scanning less than 0.02% of the Claims History Table for satisfying the query.

Calculating the Multi-Level Partition

How does Teradata Database determine exactly which partitions to look at to fulfill a request?

Multi-level partitioning is rewritten internally as single-level partitioning, and the Parsing Engine generates a combined partition number.

$(p_1 - 1) * dd_1 + (p_2 - 1) * dd_2 + ... + (p_{n-1} - 1) * dd_{n-1} + p_n$

where n is the number of partitioning expressions
  p_i is the value of the partitioning expression for level i
  d_i is the number of partitions for level i
  dd_i is the product of d_{i+1} through d_n
  $dd = d_1 * d_2 * ... * d_n <= 65535$
  dd is the total number of combined partitions

Figure 29: Partition # Formula

For this example, January 2011 is the 109th first level partition. Vermont has the 45th StateId, which is the 45th partition for the second level partition. Plugging into the formula above (Figure 29), we take 109 -1 * 75 (number of second level partitions) + the 45th partition which represents Vermont, that equates to 8145, the logical partition number for claims in Vermont for January 2011.

$$(109 - 1) * 75 + 45 = 8145$$

The Parsing Engine sends a message through the Teradata BYNET® to the AMPs directing them to look at partition #8145. Since Vermont has fewer claims than other larger states, only one partition is probed. To access the data, a total of 8,999 partitions are eliminated from the existing 9,000 partitions. This illustrates partition elimination at multiple levels.

CHARACTER Partitioned Primary Index

Teradata Release 13.10 has the ability to partition character type data using Range_N and Case_N expressions. Character type partitioning includes CHAR, VARCHAR, GRAPHIC, and VARGRAPHIC where comparisons may involve a predicate (=, >, <, >=, <=, <>, BETWEEN, LIKE) or a string function.

Typically a Teradata Session runs in a non-case-sensitive environment. However, we can turn on case sensitivity either at the table level or the session level. If it is turned on, then the partitioning expression is built with case level sensitivity.

Collation and case sensitivity considerations:

- **The session collation in effect when the character PPI is created determines the ordering of data used to evaluate the partitioning expression.**

- **The ascending order of ranges in a character PPI RANGE_N expression is defined by the session collation in effect when the PPI is created or altered, as well as the case sensitivity of the column or expression in the test value.**

- **The default case sensitivity of character data for the session transaction semantics in effect when the PPI is created will also determine case sensitivity of comparison unless overridden with an explicit CAST to a specific case sensitivity.**

Figure 30: Collation and Case Sensitivity Considerations

The examples shown here were done in a non-case sensitive environment.

Look at the ClaimPPI Table in Figure 31. It has a partition on StateCode.

Now, divide the states into ranges at a single partition level. The Primary Index is ClaimId, and we partition by RANGE_N on the StateCode, with a range of state codes. 'AA' through 'CT' would be the first partition. AA stands for Armed Forces of America and CT for Connecticut. That is the first partition range. The second partition ranges from 'DC' through 'HI' (Hawaii). Next partition is 'IA' and 'MT' (Iowa through Montana). We actually define five ranges of state code partitions and include a No Range, to support unknown state codes like 'ZZ'. The unknown codes are stored in the NO RANGE partition.

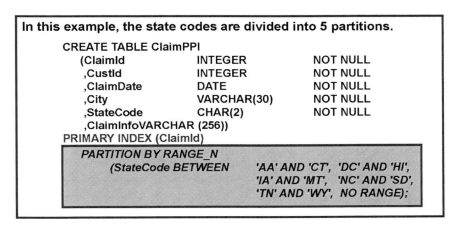

Figure 31: Claim Table with One Level of Partitioning

With the partitions defined, Teradata Database can use partition elimination. If we select a StateCode of 'OH', which is Ohio, we only look in the specific partition for the Ohio rows. We eliminate all the other partitions.

Access to the Claim Table for a StateCode that begins with the letter "O" (LIKE 'O%'), accesses one partition of the table. Data is returned for all the states whose name begins with "O", such as Ohio and Oklahoma. Additionally, we can search on ranges of states, for instance, the states between Georgia and Massachusetts.

```
SELECT * FROM ClaimPPI  WHERE StateCode = 'OH';
SELECT * FROM ClaimPPI  WHERE StateCode LIKE 'O%';
SELECT * FROM ClaimPPI  WHERE StateCode BETWEEN 'GA' and 'MA';
SELECT * FROM ClaimPPI  WHERE StateCode BETWEEN 'ga' and 'ma';
```

Figure 32: Queries that will benefit from Partitioning

Because we created the table without case sensitivity, we get the same result whether we use upper case 'GA' and 'MA' or lower case 'ga' and 'ma' characters.

Extending the example in Figure 33, we first partition first by ClaimDate. The date range starts at the beginning of 2004 and ends at the end of 2011. Each interval is one month including a NO RANGE partition. Now, subdivide the ClaimDate into StateCode and City. We take the number of partitions at the top level and multiply by the number of partitions at the sublevels. Using the full range of years starting at 2002 through 2011, causes us to exceed the total possible number of 65,535 partitions.

```
CREATE TABLE Claim_MLPPI2
    (ClaimId              INTEGER           NOT NULL,
     CustId               INTEGER           NOT NULL,
     ClaimDate            DATE              NOT NULL,
     City                 VARCHAR(30)       NOT NULL,
     StateCode            CHAR(2)           NOT NULL,
     ClaimInfo VARCHAR (256))
PRIMARY INDEX (ClaimId)
PARTITION BY
( RANGE_N
    (ClaimDate BETWEEN    DATE '2004-01-01' and DATE '2011-12-31'
                          EACH INTERVAL '1' MONTH, NO RANGE),
RANGE_N
    (StateCode BETWEEN    'A', 'D', 'I', 'N', 'T' AND 'ZZ', NO RANGE),
RANGE_N
    (City  BETWEEN        'A', 'C', 'E', 'G', 'I', 'K', 'M', 'O',
                          'Q', 'S', 'U', 'W' AND 'ZZ', NO RANGE) );
```

Figure 33: Claim Table with Three Levels of Partitioning

This example shows eight years of data that is subdivided into StateCode. This time the state codes are defined with an 'A', 'D', 'I', 'N', 'T', 'ZZ', and NO RANGE. The 'A' partition contains all the states that begin with 'A', 'B', or 'C'. The second has all the states that start with 'D' through 'H' and so on. For the last defined partition 'ZZ', include an AND in the definition, making the last state range between 'T' and 'ZZ'. If there are any states that do not fall into any of the ranges, they go into the NO RANGE partition.

At the next level, we subdivide the states into cities. The city partitions are also defined by ranges. Cities that start with 'A' or 'B', then 'C' or 'D' on down the alphabet. The last range is from 'W' to 'ZZ' with an additional NO RANGE partition for any cities that might fall outside the defined partition ranges.

The advantage of doing this is we have the ability to execute queries that look for specific states, and specific cities for a specific month, or look for a specific month for all states and all cities. There are numerous options that we can now use to take advantage of partition elimination.

```
SELECT    *
FROM      ClaimMLPPI2
WHERE     StateCode = 'GA'
AND       City LIKE 'a%';
```

Figure 34: Partition Elimination Query

We access the Claim Table where StateCode is Georgia and the city starts with an 'a%'. That includes cities such as Atlanta and Athens. This query looks at all of the months, but does not have to look at all of the partitions. We look at selected sub partitions for just Georgia and within Georgia, for specific cities that begin with the letter 'a'.

In the next query, we look for a specific month and cities that start with the letter 'a'. We look at all of the state partitions, and we retrieve the cities that begin with 'a'.

```
SELECT  *
FROM    ClaimMLPPI2
WHERE   ClaimDate = '2011-08-24'
AND     City LIKE 'a%';
```

Figure 35: Multi-Level Partition Elimination Query

Partition elimination can occur at a number of levels. Using a Character PPI also eliminates the need to have a separate lookup table, which we described in a previous example.

Changing Multi-Level Partition Definitions

Once a Multi-Level Partition Primary Index has been created, can we change the partitions?

The only level where the number of partitions can be altered is at the top level, assuming we still have partitions left. The current limit is 65,535 total partitions, where the total number of partitions is the product of all the partitions at each level. If there are 151 partitions on level one, 16 on level two, and 27 on level three, the total is 65,232 partitions. Adding one more partition would exceed the total limit of 65,535.

Surprisingly, we can ALTER the lower level partitions too. We must keep the number of partitions constant for the level we are changing, but we can actually change the partition. For example, we can change from a StoreID partition to a DepartmentID Partition. As long as the number of partitions for the level does not change, the partitioning column can be changed. The "secret" is to ALTER the table adding and dropping the partitions at the same time.

Multi-Level Partitioned Primary Index Rules Summary

Everything that applies to Single Level Character Partitioned Primary Indexes applies to Multi-Level Character Partitioned Primary Indexes. If we have more than one partitioning expression, each partitioning expression must consist solely of either a RANGE_N or a CASE_N function.

If there is more than one partitioning expression specified by a PARTITION BY clause, the product of the number of partitions defined in each of the partitioning expressions cannot exceed 65,535.

The maximum number of partitioning expressions or levels is 15. Remember to multiply the number of partitions for each of the levels together. Partitioning expressions may not contain system derived columns for the available partitions which are internally labeled PARTITION#L1 through PARTITION#L15.

The more partitions that can be eliminated, the better the query performs. Performance is proportional to the number of partitions that must be probed before the desired row is found.

Primary Index to Primary Index joins perform better when the tables are partitioned identically, and RowKey based joins are generally better than joins based on other column criteria. Coarser granularity of the PPI partitions is likely to be superior to finer partition granularity, be careful not to over partition. Partition elimination often helps join optimization, as always, use the EXPLAIN to verify your design.

For Multi-Level Partitioned Primary Indexes, Teradata Database supports base tables, Global Temporary Tables, Volatile Temporary Tables, and Non-compressed Joined Indexes. Partitioning is not supported on Hash Indexes, Compressed Join Indexes, or Queue Tables.

When analyzing the partitioning expressions in PPI tables, there are many factors to consider, these are a few of them:

- Which partitioning expression best supports the workload, CASE_N, RANGE_N, or some other expression?

- Will a NO RANGE, NO RANGE OR UNKNOWN, or UNKNOWN partition be useful if partitions are dropped?

- How many levels of partitioning are appropriate for the workload to achieve optimal partition elimination?

- Does partitioning provide performance benefits similar to a Secondary Index, and can that Secondary Index be dropped? Dropping a Secondary Index means there is no subtable to maintain and there is considerable disk savings.

- How is performance affected?

- Do queries specify equality or non-equality conditions on the primary index or the partitioning columns?

- What is the frequency for maintaining, adding, and dropping partitions?

Chapter Five
SECONDARY INDEXES

Teradata Database also provides us with Secondary Indexes. The purpose of a Secondary Index is to provide an alternate access path to the rows of a table. Secondary Indexes differ from Primary Indexes in that Secondary Indexes do not affect row distribution. That is handled by Primary Indexes that determine which AMP a row will be stored on. Secondary Indexes are purely for accessing data. The original Teradata systems came with Unique (USI) and Non-Unique (NUSI) Secondary Indexes.

Secondary Indexes add overhead. They take up additional disk space and there is some maintenance, albeit, mostly internal by the Teradata Database. Secondary Indexes can be added or dropped dynamically as needed. Of course, we would not necessarily want to do that in the middle of the day when users are accessing the data and other processing is going on. Use some discretion when deciding when to add and drop Secondary Indexes, but know these indexes can be added and dropped at will.

- **Do not affect table distribution.**
- **Do add overhead, both in terms of disk space and maintenance.**
- **May be added or dropped dynamically as needed.**
- **Are chosen to improve access performance.**

Figure 36: Secondary Index Rules

Secondary Indexes improve performance on the system. A Secondary Index can be chosen as an alternate access path, or in the case of a Unique Secondary Index, to maintain uniqueness of a data column.

Unique Secondary Indexes

Figure 38: Access via a USI, shows the Car Table that is distributed on the basis of the Non-Unique Primary Index (Phone column). We refer to Car as a base table. To maintain the uniqueness of the Primary Key from the Logical Data Model, we add a Unique Secondary Index on the Ownr column which, in turn, creates a Unique Secondary Index Subtable. This immediately causes space to be used and that is the initial overhead for creating the index.

Create USI Syntax

> **CREATE UNIQUE INDEX (Ownr) on Car;**

Figure 37: Create a Unique Secondary Index

The index subtable row is created by taking information from the base table row. Let's use the row for phone number 555-7777 on AMP4 as an example. The Ownr value is 45 and the Color is Green. To create the Unique Secondary Index (USI), the values of the column are hashed in the same way Primary Index values are hashed. The Hash Code is used by the Teradata BYNET® to determine where to store the USI subtable row (index value, hash value, hash bucket, AMP number). The base table row is stored on AMP4 based on the Primary Index and the subtable row for value 45 is stored on AMP2.

Unique Secondary Index (USI) Access

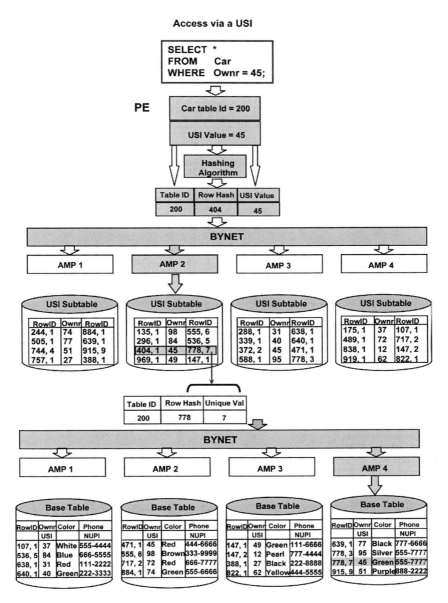

Figure 38: Access via a USI

Teradata® Database Index Essentials

The base table row includes the hash value and the uniqueness value that is called the RowID. The subtable row also has a hash value and a uniqueness value that comprise the RowID for the USI subtable row. The USI row also includes the index value which is 45 in this example.

Finally, the USI subtable row contains the RowID from the base table row. It is by that RowID that the system can find the base table row when it only has the USI value.

When we access via the USI, in this case selecting all the columns from the Customer Table, the optimizer has a value that can be hashed.

The three pieces of information needed by the Teradata system to find a row in a table via an index are the TableId, the RowHash, and the Index Value. In the example, we have customer TableId 200, USI Value 45, and the RowHash 404 which was generated in the Parsing Engine by the optimizer.

This information is sent to the Teradata BYNET®. Using its maps, the Teradata BYNET® "knows" from the Hash Code that the row will be found on AMP2.

The request is sent to AMP2 and the USI subtable row is accessed. They system then verifies the Hash Code and the USI value, 45, and obtains the RowID for the base table row. The message is then sent to the Teradata BYNET® where the RowHash is used to verify which AMP the base table row will be located on. The message is the sent via the Teradata BYNET® to AMP4.

Once the message is received by AMP4, the File System is used to locate the row in the base table. This is an example of a two-AMP operation.

USI access is always considered to be a two-AMP operation. We go to the first AMP using the RowHash of the USI. From the USI we get the RowID for the base table, and we go to the second AMP to retrieve the base table row.

Non-Unique Secondary Indexes

Non-Unique Secondary Indexes are slightly different than UPIs, NUPIs, and USIs

What we have seen so far is that for Primary and Unique Secondary Indexes we put their values into the Hashing Algorithm and get a 32-bit Row Hash. Using the TERADATA BYNET's® map, rows are stored "globally" on the AMPs. In other words, rows are randomly distributed across the AMPs based in the value of the row hash generated by using the primary index value.

Non-Unique Secondary Index values are also hashed in the same manner. The only difference is Non-Unique Secondary Index rows are stored on the same AMP as the base table row.

CREATE Non-Unique Secondary Index Syntax

```
CREATE INDEX (Color) on Car;
```

Figure 39: Create a Non-Unique Secondary Index

When we create a Non-Unique Secondary Index (NUSI) on the Color column from our Car Table, the system creates a non-unique secondary index subtable. We indicate the column name and the table it is from. A NUSI subtable is created on all the AMPs. The NUSI distinction is that we take the name column data, we hash it, but we store the rows locally on the same AMP as the base table rows. Hence the distinction "local" hash versus the UPI, NUPI, and USI "global" hash.

Teradata® Database Index Essentials

Non-Unique Secondary Index (NUSI) Access

Let's SELECT all the columns from the Car Table where the Color is Red.

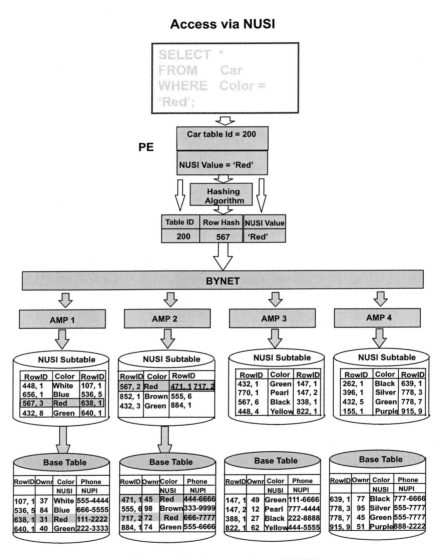

Figure 40: Access via a NUSI

A Car Table generally has multiple vehicles whose color is red and red cars may exist on every AMP because the data was distributed by the NUPI on phone number, rather than on car color.

Again, to access data via an index, Teradata Database requires the TableId which is 200 for this example; the RowHash of the NUSI which is 567, and the actual NUSI value which is "Red".

Using the NUSI is an All-AMP operation because we have to look at the NUSI subtable on every AMP. The NUSI subtable tells us there is one "Red" car on AMP1 and two "Red" cars on AMP2. AMPs 3 and 4 have no "Red". The subtable rows have the RowID (RowHash of "Red" plus the uniqueness value counter), the NUSI value, and the RowID from the base table which is the unique identifier of the row in the base table.

The optimizer has the option to use or not use the NUSI based on the selectivity of the NUSI values in the subtable. It "knows" how much data has to be read, how many I/Os it needs to do and how many data blocks must be read based on Collected Statistics.

A NUSI is useful when the system does not have to read every data block and that depends upon the percentage of the number of rows that qualify and the number of rows in the data blocks. We have the table with the data blocks and the NUSI only has to go to a hand full of the AMPs, therefore it will be beneficial to use the NUSI. If, on the other hand, there are qualifying rows on every AMP and possibly in every block, then the optimizer may not use the NUSI, but instead do a Full Table Scan. The optimizer chooses the best access method. There is no user override or "hints" that can be given to the optimizer, it uses Statistics or Random AMP samples to make its determination.

The EXPLAIN output shows us what the optimizer chose. It says "by way of index #..." or "we do a full table scan...". For a NUSI to be considered the optimizer must have Statistics, otherwise it is pretty much a guarantee that the NUSI will not be used. If we expect the NUSI to be chosen by the optimizer, we must collect Statistics.

```
ORDER BY              ColumnName;
ORDER BY   HASH       (ColumnName);
ORDER BY   VALUES     (ColumnName);
```

Figure 41: NUSI ORDER BY Options

There are three ordering options for NUSIs. ORDER BY ColumnName defaults to ordering by the hash of the column value. ORDER BY HASH (ColumnName) stores the rows in column value hash sequence and is good for data value searches. ORDER BY VALUES (ColumnName) sorts on the column value and is good for range searches

Additional NUSI Criteria

Rows Per Block

NUSI usage depends on the number of rows per block that qualify. For example, if more than one row per block qualifies, then a Full Table Scan of the base table will be much faster than NUSI access.

Assume 100 rows per block and that 1% of the data qualifies. Every block is read, so the Full Table Scan is faster.

By contrast, if fewer than one data row per block qualifies, then the optimizer uses the NUSI to access the data because it is faster than a Full Table Scan.

If there were 100 rows per block and 1 out of every 1,000 rows qualify, then one in every 10 blocks is read, and NUSI access is faster.

When present, Statistics are used by the optimizer to makes its choice. Without Statistics, the optimizer does not even consider using the NUSI.

Data Distribution

Another consideration is the unevenness of data value distribution. This occurs when some values represent a large percentage of the table rows and other values have fewer rows

In the Car Table, for example, we want to have a NUSI on car color. If we specify red cars, chances are that the optimizer would not use the NUSI because it is one of the most popular car colors and there are a lot of red cars. The optimizer would instead do a Full Table Scan. If we chose the purple cars, there would not be so many, and the optimizer would have a greater propensity to use the NUSI. NUSIs are good for values that represent a small percentage of the table.

Another example is phone service for a large corporation. They make 100,000 calls per month. For monthly call information, we're likely to do a Full Table Scan. For residential phone customers who make an average of 20 calls per month, that would probably be a good NUSI. Assuming we collected Statistics, the optimizer would be likely to use the NUSI for those values. The caveat to this is that instead of a NUSI, this might be a great candidate for a Sparse Single Table Join Index (special case of a STJI that is qualified with a WHERE clause - which will be covered in the next chapter).

Covered Indexes

A covered index is when a query can be satisfied using an index subtable without having to go to the base table to satisfy the query. A NUSI can be used as a Covered Index. For example, let's say we have a table that has 200 columns associated with it, and the vast majority of our queries only use 20 of the 200 columns. We can build a NUSI that just includes those 20 columns. Any queries that use a subset of those 20 columns could then use the NUSI. We may do a Full Table Scan of the table to find the answer set, but it would scan the much smaller NUSI subtable, rather than the base table.

When building a NUSI, and for this example assume multiple columns, we can order by hash or order by values where the hash would allow quality access on a specific column, while covering the query. One could use that NUSI to cover the query. Order by values also gives us some range access capabilities.

An advantage of ORDER BY HASH is that the data in the NUSI subtable is in hash sequence. That NUSI is valuable for covering queries and can be considered in join plans. Teradata Database does not join to value ordered NUSIs.

- **The optimizer considers using a NUSI subtable to "cover" any query that references only columns defined in a given NUSI.**
- **These columns can be specified anywhere in the query including:**
  - SELECT list
  - WHERE clause
  - Aggregate functions
  - GROUP BY clauses
- **Presence of a WHERE condition on each indexed column is not a prerequisite for using the index to cover the query.**
- **NUSIs (especially a covering NUSI) are *considered by the Optimizer in join plans* and can be joined to other tables in the system.**

Figure 42: NUSI Covering Criteria

Here are a few examples:

NUSI considered for index covering:

```
CREATE INDEX IdxOrd
        (o_OrderKey
        , o_Date
        , o_TotalPrice)
ON Orders ;
```

Figure 43: Three column NUSI

We created a NUSI on OrderKey, OrderDate, and TotalPrice. We do not order by hash or order by values, so the default is to order by hash. Any query that uses a combination of those three columns fully specified, uses the NUSI to find the row. If we specify a subset of those columns a NUSI subtable scan can used.

NUSI considered for index covering and ordering:

```
CREATE INDEX IdxOrd2
        (o_OrderKey
        , o_Date
        , o_TotalPrice)
ORDER BY VALUES (o_OrderKey)
ON Orders ;
```

Figure 44: Value Ordered NUSI

For this example, we have an index with an ORDER BY VALUES. This gives us range access on OrderKey for that specific table.

Query to access the table via the OrderKey:

```
SELECT      o_Date
            , AVG(o_TotalPrice)
FROM        Orders
WHERE       o_OrderKey >1000
GROUP BY    o_Date ;
```

Figure 45: Range Constrained Query

For this sample query, we select OrderDate and Average of TotalPrice from the Orders Table. The two columns we reference are both of the indexes that we created in the first two examples. Either is chosen by the optimizer. When we specify an OrderKey greater than 1,000, we have a range type constraint. The second example, ORDER BY VALUES, is a little more efficient because we do not scan the entire Orders base table. We only look at the NUSI subtable for orders that are greater than 1,000.

Let's look at another example.

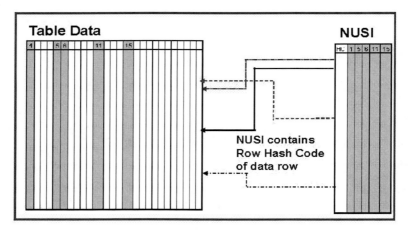

Figure 46: NUSI Index Covering

We have a table that has 26 data columns and on it we create a NUSI that uses 5 columns. The NUSI contains the RowHash code of the data row. If we can satisfy the query using the NUSI, the physical I/O savings can be considerable. The savings is based on the number of bytes (columns) in NUSI definition versus number of bytes (columns) in the base table definition. For this example, the savings were between 60-80 percent. The query was satisfied with NUSI access only, no base table access was necessary.

NUSI on PI Columns of PPI Tables

The NUSI can be defined on the same column(s) as the Primary Index of a PPI table. For a given value, accessing the ***NUSI on the PI of the PPI table results in a single AMP operation.*** Let's say we access insurance policies that are paid bi-annually. We can put a PPI on policy with a partition on the date. Then put a NUSI on policy. The NUSI accesses only the partitions for the month when the policy was paid.

This also works in a retail environment when we want to access seasonal items sold in a store. If we had a PPI on Store and Item with a partition on Date and a NUSI on Item, then the NUSI accesses only the partitions for the months when an item was sold.

Place a USI on NUPI

Now let's revisit the earlier example where we had a Claim Table and where we partitioned by month. For a query where we select from the Claim Table for specific ClaimId (260221), using only a primary index, must look in every one of the partitions. We had 120 partitions and the EXPLAIN cost was 0.09 seconds. We had to do additional work because we had to look in all the partitions. One of the options that we can do is create a Unique Secondary Index on ClaimId.

```
CREATE UNIQUE INDEX (ClaimId)
ON ClaimPPI;
```

Figure 47: Create a USI on a PPI table

We created a Unique Index because we know ClaimId truly is unique in this table.

```
SELECT  *
FROM    ClaimPPI
WHERE   ClaimId = 260221;
```

Figure 48: Select on USI Column

Remember that when the Unique Secondary Index is created, a subtable is built. The subtable row has the partition number, the RowHash, and the uniqueness value of the base table row.

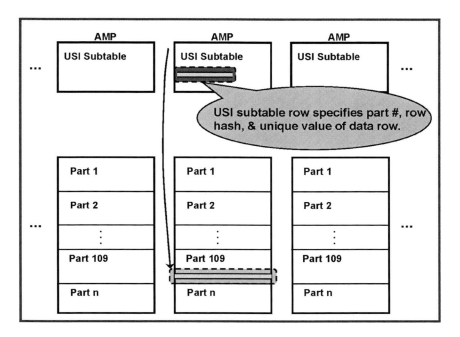

Figure 49: Create a USI on a Non-Unique PPI Table

The partition number, the RowHash, and the uniqueness value, together total 10 bytes. Two bytes for the partition number, four bytes for the hash, and four bytes for the uniqueness value. Because we have that information in the RowID of the subtable row, we can go directly to that data row in the base table.

This example is a two-AMP operation. We go to the USI subtable, pick up the subtable row which contains the RowID of the base table, and that refers us to the base table row on another AMP.

This usage of a USI eliminates partition probing and uses row-hash locks. It only works this way if the values in PI column(s) are unique (even though we defined a NUPI). The USI then maintains the data uniqueness.

Placing a USI on a NUPI is only supported on PPI tables.

Place a NUSI on NUPI

We can also place a NUSI on the ClaimId column. If the ClaimId was not unique, we could build a NUSI on it.

CREATE INDEX (ClaimId) ON ClaimPPI;

NUSIs are generally All-AMP operations.

Figure 50: Select on NUSI Column

A NUSI it doesn't mean a Full Table Scan, but it usually means All-AMPs.

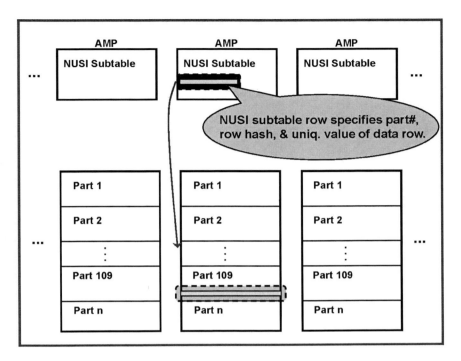

Figure 51: Create a NUSI on a Non-Unique PPI Table

This scenario uses a NUSI that results in a single AMP operation. The Parsing Engine knows this. By creating a Secondary Index on ClaimId for this particular table, we're selecting from the Claim Table where the ClaimId is 260221. We go to one AMP, pick up the NUSI subtable row. This example gives us one RowID that points us to the base table row. If we have multiple claims for that same ClaimId, we pick up the multiple claims. The optimizer generates a plan for single-AMP NUSI access with row-hash locking (instead of table-level locking).

This use of NUSIs eliminates partition probing. These One-AMP operations use row-hash locking. The NUSIs can be used with unique or non-unique PI columns and must be accessed by equality conditions. NUSI Single-AMP operations are only supported on PPI tables and are allowed when using MultiLoad to load a table.

This demonstrates that a USI or a NUSI can be placed on the Primary Index of a PPI table and we can use that NUSI or USI for value access.

Secondary Index Review

Let's review Secondary Indexes. We have demonstrated that Secondary Indexes provide faster set selection. They may be Unique (USI) or Non-Unique (NUSI). If we choose a NUPI, a USI may be used to maintain uniqueness of a column or the Primary Key column.

The system maintains a separate subtable for each Secondary Index. If we choose to have five Secondary Indexes on a table, there will be five Secondary Index Subtables associated with that base table that have to be maintained by the Teradata system.

Secondary Indexes are optional and can be chosen to significantly reduce base table I/O during value access and join operations.

For Non-Unique Secondary Indexes, consider how they support Nested and Merge Joins operations.

The optimizer can choose to use bit-mapping for weakly selective NUSIs and alleviate limitations associated with composite NUSIs. The optimizer uses a bit mapping technique to combine multiple weakly selective NUSIs to create an array of RowIDs that are used to retrieve base table rows. NUSI bit mapping is a fast way to access data using NUSIs that individually would not be selective enough for the optimizer to use them. The optimizer only uses NUSI bit mapping when the weakly selective NUSIs are ANDed together. There is a tremendous cost saving associated with NUSI bit mapping.

NUSIs can be used to "cover" a query so we can reduce or avoid base table access. When choosing a Non-Unique Secondary Index, use caution on choosing columns with highly volatile data values because both base table and subtable rows must be changed when the data values change. A separate subtable must be built and maintained and Statistics must be collected. There is a lot to be gained by having Secondary Indexes. The EXPLAIN query plan shows whether the optimizer has chosen to use the Secondary Index or not.

Secondary Indexes are secondary access paths to the data. USIs ensure uniqueness of data and require two AMPs for access. NUSIs require all AMPs for access, they may support some join operations, and they used based on data selectivity.

Chapter Six
JOIN INDEXES

Primary and Secondary Indexes are chosen based on known workloads. As new applications are integrated into the data warehouse, sometimes the already established table indexes are not the most suitable for the new workloads. It would be inappropriate to create a second or duplicate table using a different PI, because the two tables take up twice the space, multiple tables are maintained, and it is possible to introduce data anomalies. Join Indexes (JIs) are an appropriate alternative.

There are several types of Join Indexes:

- Single-Table
- Multiple-Table
- Sparse
- Global
- Value ordered
- Combinations of the above.

A JI is a separate object within a Database or User with its own Primary Index, uses permanent storage space, and can be Fallback protected. The choice of PI for the JI is determined by the desired access for the workload it is intended to enhance. JIs give the optimizer additional options for access, especially when the base table Primary and Secondary Indexes do not provide usable access paths. Unlike Secondary Indexes, Join Indexes are not subtables associated with base tables and do not contain base table RowIDs.

Join Indexes are created to improve join performance and optimize table and join access. Join Indexes can specify multiple columns from one or more tables. When JIs specify columns from multiple tables, they actually pre-join the data.

Two processes that can require significant resources, because of a large numbers of rows, are joins and aggregation. Join Indexes optimize both of these processes when the JI table contains the joined and aggregated information, greatly reducing processing time. In fact, Join Indexes can eliminate join processing altogether. Other RDBMSs refer to a Join Index as a materialized view.

A specialized type of row compression, that reduces storage within the Join Index table, is also available. We will cover that topic in the Compressed Join Index section.

Aggregate Join Indexes are an alternative to summary tables. They have the advantage of being automatically updated, whereas summary tables require separate and constant update processing.

SINGLE TABLE VS. MULTI-TABLE JOIN INDEX OVERVIEW

Single Table Join Indexes (STJIs) are a special case of Join Index. They frequently have their Primary Index on a different column than the PI of the base table. Quite often, they are an excellent alternative for new applications when data values of the base table Primary Index are unknown, and therefore not helpful for accessing data. When the optimizer uses this type of STJI, it gives the advantage of having PI access on the Join Index. It also avoids using base table Secondary Index access or base table Full Table Scans to access the data. STJIs can also eliminate redistribution for joins between the STJI table and other tables, if the Primary Indexes of the tables to be joined are the same.

Multiple Table Join Indexes (MTJIs) create a pre-joined table that is stored on the AMPS. This table can contain data from all of the pre-joined tables, as well as aggregated data, and may or may not contain the actual join columns from the base tables.

OPTIONS FOR JOIN INDEXES

Sparse Join Index

A Sparse Join Index includes a WHERE clause to limit the rows that are stored in the Join Index. The Sparse Join Index takes up less space, is faster to create, and more efficient to access than a JI where row selection is unlimited. The Sparse STJI is useful when we need to access or join to only a portion of a table or set of tables.

Global Join Index

A Global Join Index includes the RowID of the base table row as part of the data contained in the Join Index. Global JIs are useful for partial index searches. They qualify the rows with a Full Table Scan of the Global Join Index, which even though it is a FTS, it is done on much smaller table than the base table and is more efficient. We then use the RowID value of the base table within the Global JI to "join back" to the base table. This saves significant I/O and processing time.

Join Index Considerations

Join Indexes can contain up to 64 columns from their referenced base table(s). The Join Index includes columns that are commonly used by queries and workloads to eliminate or minimize the need to access the base table(s).

A Join Index is a database object and acts like a separate table, although it cannot be accessed directly. SQL always specifies data from the base tables. It is up to the optimizer to choose to use the JI, if it offers the most efficient access.

Join Indexes can have their own Secondary Indexes and Collecting Statistics is recommended on JIs. Join Indexes can also be partitioned, unless they are compressed (more about this later). Additionally, Join Indexes can be defined for base tables with triggers.

Join Index Example – Customer and Orders Tables

For the Join Index examples in this section, we will use the Customer and Orders Tables. They are two small base tables.

```
CREATE SET TABLE Customer
    (c_CustId      INTEGER NOT NULL
    ,c_LName       VARCHAR(15)
    ,c_FName       VARCHAR(10)
    ,c_Address     VARCHAR(50)
    ,c_City        VARCHAR(20)
    ,c_State       CHAR(2)
    ,c_ZipCode     INTEGER)
UNIQUE PRIMARY INDEX ( c_CustId );
```

Figure 52: Customer Table

Production tables generally have 70 or 90 columns in them. Our Customer Table has seven columns and the Orders Table has nine columns. Often production tables have many millions of rows, here we have 72,000 orders and 5,000 customers.

```
CREATE SET TABLE Orders
   ( o_OrderId        INTEGER NOT NULL
    ,o_CustId         INTEGER NOT NULL
    ,o_OrderStatus    CHAR(1)
    ,o_TotalPrice     DECIMAL(9,2) NOT NULL
    ,o_OrderDate      DATE
                      FORMAT 'YYYY-MM-DD' NOT NULL
    ,o_OrderPriority  SMALLINT
    ,o_Cclerk         CHAR(16)
    ,o_ShipPriority   SMALLINT
    ,o_Comment        VARCHAR(79))
UNIQUE PRIMARY INDEX ( o_OrderId );
```

Figure 53: Orders Table

The Primary Index of the Customer Table is CustId and the Primary Index of the Orders Table is OrderId. Normally, the tables are joined on the CustId column.

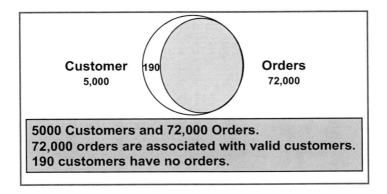

Figure 54: Customers with Orders

All of the orders have valid customer numbers, but there are 190 customers which have no orders.

Teradata® Database Index Essentials

Compressed Join Index

Keeping the Customer Table (Figure 52) and Orders Table (Figure 53) in mind, we are now ready to delve into Compressed Join Indexes.

A Compressed Join Index is "row compressed". For rows where multiple columns have the same data, that data is stored only once. For example, we see a Compressed Join Index in Figure 55 below. CustID and LName are stored only once for each customer, regardless of how many orders are present.

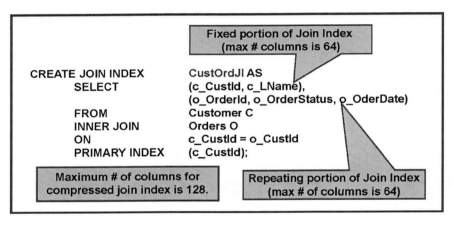

Figure 55: Compressed Multi-Table Join Index Syntax

This Join Index includes columns from multiple tables. The syntax includes a SELECT statement with a join to specify the multiple tables, Customer and Orders. The compress syntax uses two sets of parenthesis. The first set is around the "fixed, non-repeating portion". The second set of parenthesis is around the "repeating portion". The parenthesis around (c_CustId, c_LName) indicates the fixed portion of the data. The parenthesis around (o_OrderId, o_OrderStatus, o_OrderDate) represents the data that repeats for c_CustId and c_LName.

If a customer has multiple orders, we can use the compressed format. Row compression, however, does not support multiple customers who share the same orders.

JI row compression permits up to 64 columns in the fixed portion and another 64 columns for the repeating portion. In Figure 56, there is a row for a customer named Grills. Grills has three orders that are completed, and one order that remains open.

Fixed Portion		Variable Portion			
c_CustId	c_LName	o_OrderId	o_OrderStatus	o_OrderDate	
1443	Grills	102292	C	2008-11-30	Within the join index, this is one row of data representing the orders for a customer.
		135893	C	2009-11-30	
		157093	O	2010-10-16	
		135993	C	2009-12-14	
4000	Wolfe	142000	C	2009-12-20	One row in Join Index.
		149600	C	2009-12-29	
		154798	C	2010-04-17	
5809	Thatcher	149698	C	2009-12-31	One row in Join Index.
		156599	C	2010-10-05	
		101199	C	2008-09-11	
		158999	O	2010-10-23	
		152399	C	2010-06-30	

Figure 56: Compressed Multi-Table Join Index Query Results

Inside the Join Index, the CustId and LName are not repeated in multiple rows, the data is stored a single row. The output report displays all four rows for Customer Grills. Likewise, Wolfe is internally represented in a single row with three orders and Thatcher is stored as a single row with five orders.

The Primary Index of our JI table is CustId, so the data is distributed across all the AMPs, based on the CustId.

Non-Compressed Multi-Table Join Index

Let's contrast this to a Non-Compressed Join Index.

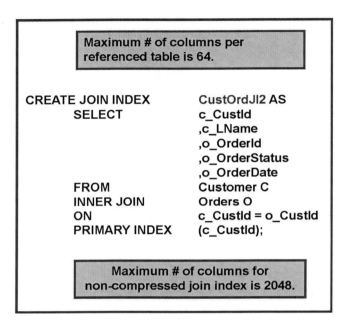

Figure 57: Non-Compressed Multi-Table Join Index Syntax

The syntax is similar, but there are no parentheses. This syntax allows multiple customers to share multiple orders. Partitioning on non-compressed Multi-Table Join Indexes is also available and allows as for many as 2,048 columns across the set of tables. In this example, Figure 58, Kruglak has an order in common with Suffren, so each order for each customer must have its own row in the Non-Compressed Join Index.

Fixed Portion		Variable Portion		
c_CustId	c_LName	o_OrderId	o_OrderStatus	o_OrderDate
1444	Kruglak	102293	C	2008-11-30
1444	Kruglak	135894	C	2009-11-30
1444	Kruglak	157095	O	2010-10-16
1444	Kruglak	135996	C	2010-04-17
5000	Suffren	135996	C	2010-04-17
5000	Suffren	142001	C	2009-12-20
5000	Suffren	149601	C	2009-12-29
5812	Schneider	149699	C	2009-12-31
5812	Schneider	156592	C	2010-10-05
5812	Schneider	101192	C	2008-09-11
5812	Schneider	158992	O	2010-10-23
5812	Schneider	152392	C	2010-06-30

Within the join index, each order is represented by a separate join index row.

Common orders between Suffren and Kruglak, each represented by a separate join index row.

Figure 58: Non-Compressed Multi-Table Join Index Query Results

Because there is repeated information, the Non-Compressed Join Indexes physically take up more storage space on the disks than the Compressed Join Indexes.

Compressed and Non-Compressed Join Index Storage Comparison

In the Compressed Join Index, there are multiple orders associated with a single row.

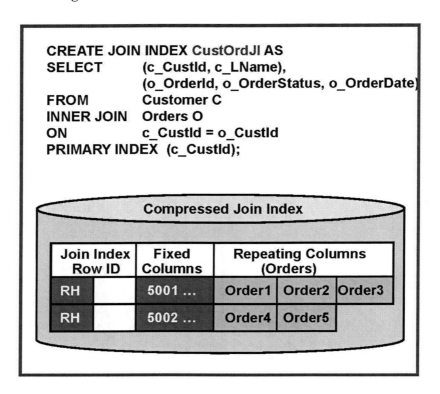

Figure 59: Compressed Join Index Storage

A Non-Compressed Join Index has separate rows for each one of the orders.

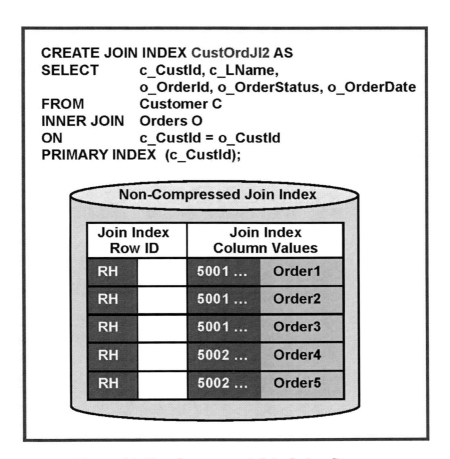

Figure 60: Non-Compressed Join Index Storage

Even in this small example, significant savings within a Compressed Join Index are apparent. The Compressed JI only uses two rows to store five orders, whereas the Non-Compressed Join Index needs five rows to store the five Orders.

```
SELECT    TableName, SUM(CurrentPerm)
FROM      DBC.TableSize
WHERE     DatabaseName = USER
GROUP BY 1 ORDER BY 1;
```

Figure 61: Query to Determine Space Usage

TableName	SUM(CurrentPerm)
CustOrdJI	1,044,480
CustOrdJI2	2,706,432

Figure 62: Join Index Space Usage Comparison

The downside of the Compressed Join Index is that it may be less usable, because partitioning is not available and many-to-many relationships are not supported.

Will a Compressed Join Index Help Access?

List the valid customers who have <u>open</u> orders

SELECT	c_CustId
	,c_LName
	,o_OrderDate
FROM	Customer C
INNER JOIN	Orders O
ON	c_CustId = o_CustId
WHERE	o_OrderStatus = 'O'
ORDER BY	1;

SQL Query	Cost
Without Join Index	.49 seconds
With Join Index	.18 seconds

Figure 63: Query with EXPLAIN Cost for Open Order Request

In this example, we select rows where OrderStatus is open.

c_CustId	c_LName	o_OrderDate
1391	Carter	2010-10-06
1906	Elmore	2010-10-22
1969	Garcia	2010-09-14
2916	Perez	2010-10-24
2954	Higa	2010-10-15
4336	Friedberg	2010-09-20
5396	Salaway	2010-10-21
:	:	:

Figure 64: Sample Output

The answer set will be the same if we use the JI or the base table, but without using the Join Index, the EXPLAIN plan cost is 0.49 and with the Join Index the cost is 0.18. All of the data to satisfy the query is available in the Join Index, and no base tables are accessed.

> **All referenced columns are part of the Join Index.**
>
> **Optimizer chooses Join Index rather than doing a join.**
>
> *Join Index covers the query and helps this query.*
>
> **The Compressed Join Index was used for this example.**

Figure 65: Optimizer Chose Join Index to Cover Query

The next example asks for customer and address information.

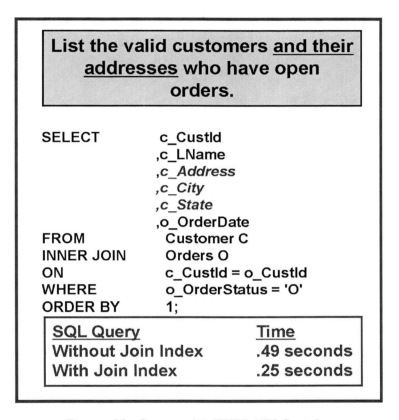

Figure 66: Query with EXPLAIN Cost for Valid Customers with `Open Orders

The JI does not completely cover this query, but Teradata's Database optimizer knows that the remaining information is in the Customer Table. Since both the base table and JI tables are hashed on CustId, the Teradata Database can join them AMP locally.

Results:

c_custid	c_LName	c_address	c_city	c_state	o_OrderDate
1391	Carter	2300 Madrina Ave	Carson City	NV	2010-10-06
1906	Elmore	903 Noessa Ave	Tampa	FL	2010-10-22
1969	Garcia	36 Main Street	New York	NY	2010-09-14
2916	Perez	4564 George Street	New Brunswick	NJ	2010-10-24
2954	Higa	1083 Beryl Ave	Los Angeles	CA	2010-10-15
4336	Friedberg	4021 Paradise Way	St. Paul	MN	2010-09-20
5396	Salaway	5603 Main Street	Pierre	SD	2010-10-21
:	:	:	:	:	:

Figure 67: Valid Customer Answer Set

This yields about a 2:1 cost difference in performance.

Without the Join Index, the cost is 0.49 because data was physically redistributed. Now we join the Join Index to the base table, and skip the redistribution step.

- **Some of the referenced columns are NOT part of the Join Index. The Join Index does not cover the query, but is used in this example.**
- *A Join Index is used in this query and is merge joined with the Customer table.*

Figure 68: Optimizer Chose Join Index for Partial Covering

We have additional flexibility when we create a Non-Compressed Join Index.

```
CREATE JOIN INDEX      Cust_Ord_JI3 AS
       SELECT          c_CustId
                       , c_LName
                       , o_orderid
                       , o_orderstatus
                       , o_OrderDate
       FROM            Customer C
       INNER JOIN      Order O
       ON              c_CustId = o_CustId
       PRIMARY INDEX   (c_CustId)
       PARTITION BY RANGE_N (o_OrderDate BETWEEN
                       DATE '2002-01-01' AND DATE '2011-12-31'
                       EACH INTERVAL '1' MONTH) ;
```

Figure 69: Create a Partitioned Join Index

This example in Figure 69 builds a Join Index between Customer and Orders. This time, the partition is on OrderDate.

We want to locate all the valid customers that have open orders for January 2011. Will partitioning help this query?

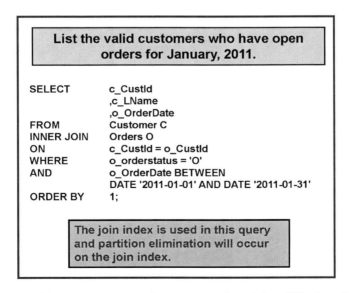

Figure 70: Using a Join Index with Partition Elimination

Yes. We can have partition elimination on both the base table and the Join Index. In this case, date range is used for partition elimination.

Teradata Database Join Index Optimization

Before Teradata Release 13, Join Indexes had to be very specific or the optimizer would not choose to use them because JIs were heuristic based instead of cost based.

Starting at Teradata Release 13.0, there is a vast difference in Join Index usage in a much wider variety of contexts. The optimizer can use information from the Join Index, even if it does not use the Join Index, to help speed up queries.

Multi-Table Join Index Usage

Multi-Table Join Indexes have various advantages, depending on their Primary Index, with respect to the base tables they will be joined to.

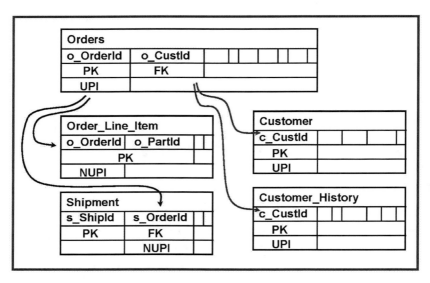

Figure 71: Columns for Multi-Table Join Index

For this example, we join the Orders and Customer Tables. If we use the Orders Table, which we commonly join to OrderLineItem and Shipment, we do not need a JI. The join is on OrderId, the Primary Index of all three base tables. We also need to join to Customer and Customer History to Orders. Consider building multiple Multi-Table Join Indexes for these processes.

One of the big costs in doing a join between Orders and Customer is redistributing the Orders Table on the hash of CustId. To avoid this redistribution, we create a JI on the Orders Table with a PI of CustID, this serves as an alternative PI on the Orders Table.

> **Without a Join Index, redistribution or duplication is required to complete the join of Order and Customer (or Customer History).**

Figure 72: Eliminate Redistribution and Duplication

We create a Single Table Join Index on the Orders Table and call it Order_JI, using the CustId as the NUPI (of the Join Index). As we insert Orders into the Orders Table, the Join Index is automatically updated.

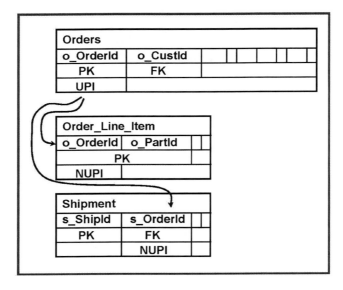

Figure 73: Join Index is Automatically Updated

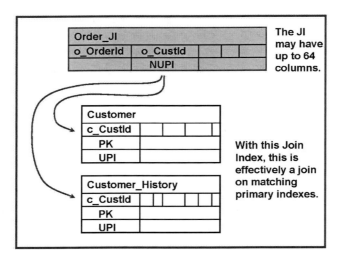

Figure 74: Order Join Index

Now, when we join Orders to Customer, instead of the optimizer having to redistribute the Orders Table, it can join Customer to Order_JI and avoids data redistribution.

Creating a Single Table Join Index

Single Table Join Indexes are common within the Teradata Database because of their flexibility.

```
CREATE JOIN INDEX  Order_JI  AS
    SELECT              (o_CustId),
                        (o_OrderId,
                        o_OrderStatus,
                        o_totalprice,
                        o_OrderDate)
    FROM                Orders
    PRIMARY INDEX       (o_CustId);
```

Figure 75: Compressed Single Table Join Index Syntax

STJI supports both compress and non-compress options. The compressed version takes up less space, but may lose flexibility because it cannot be partitioned.

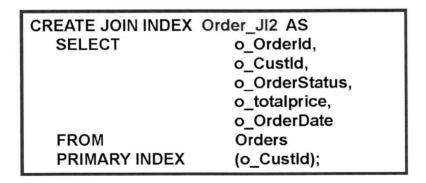

```
CREATE JOIN INDEX  Order_JI2  AS
    SELECT              o_OrderId,
                        o_CustId,
                        o_OrderStatus,
                        o_totalprice,
                        o_OrderDate
    FROM                Orders
    PRIMARY INDEX       (o_CustId);
```

Figure 76: Non-Compressed Single Table Join Index Syntax

In the previous example, Figure 76, where we create a Non-Compressed Join Index, there are no parentheses in the syntax, and also we have the option to partition on Order Date.

- The Orders base table is distributed across the AMPs based on the hash value of the o_OrderId column (primary index of base table).

- The Join Index (Order_JI) represents a subset of the Orders table (selected columns) and is distributed across the AMPs based on the hash value of the o_CustId column.

- The optimizer can use this Join Index to improve joins using the CustId to join with the Orders Table.

We do a SELECT, looking for open orders between the Customer and the Orders Tables. Let's compare optimization choices.

```
SELECT         c_CustId
               , c_LName
               , o_OrderDate
FROM           Customer C
INNER JOIN     Orders O
ON             c_CustId = o_CustId
WHERE          o_OrderStatus = 'O'
ORDER BY       1;
```

c_CustId	c_LName	o_OrderDate
1391	Carter	2010-10-06
1906	Elmore	2010-10-22
1969	Garcia	2010-09-14
2916	Perez	2010-10-24
2954	Higa	2010-10-15
4336	Friedberg	2010-09-20
5396	Salaway	2010-10-21
:	:	:

Figure 77: List Valid Customers with Open Orders

We select columns from the Customer and Orders Tables.

SQL Query	Cost
Without Join Index	.49 seconds
With Join Index	.22 seconds

- The rows of the customer table and the Join Index are located on the same AMP.
- *A single table Join Index will help this query.*

Figure 78: Query Cost Results

Without the Join Index, we join the base tables at a cost of 0.49. With the Join Index, we join the Join Index to the base table with a cost of 0.22 seconds. That is 2.5 times faster. The cost numbers are values that come from the optimizer in the EXPLAIN output and they are valuable for comparison purposes, and do not imply execution time.

Index Covering with a Single Table Join Index (STJI)

This is the 26 column table that we showed in the NUSI example. This time, we use it to illustrate the five column Single Table Join Index.

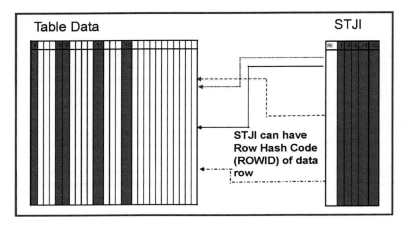

Figure 79: Vertical Partitioning with a Single Table Join Index

Queries can be satisfied with Single Table Join Index access and the Join Index is maintained automatically by the system. Additionally, the Single Table Join Index can contain the Hash Code or RowID of the data row from the base table.

Single Table Join Indexes and Non-Unique Secondary Indexes

How do we choose between a STJI and a NUSI?

These are the Similarities:

The similarities between a NUSI and a STJI are that they can both be defined with the same columns, Index Covering applies to both a STJI and a NUSI, and we have the ability to do Value Ordering on both the NUSI and the STJI.

The Differences:

The basic difference between a NUSI and a JI is that the Join Index is similar to a table with a Primary Index of its own and may have additional other columns. If we could access the JI directly, it would look like a normal data table. The Single Table Join Index row can be stored on the same AMP or a different AMP as the base table row, whereas a NUSI must always be stored on the same AMP as its base table row.

Tradeoffs:

Significant time is spent choosing Primary Indexes to get your database up and running. When new applications are added, the existing Primary Index may not be the best choice. If the new application requires a Primary Index on a different column, it is inappropriate to create a duplicate table because of maintenance, expense, and data consistency issues. Instead, create a Single Table Join Index with the desired Primary Index. That gives the new application better access to the system. From that perspective, using Single Table Join Indexes can be more beneficial than the original Primary Index or a NUSI.

MultiLoad supports NUSIs because the rows are stored AMP locally. MultiLoad is "allergic" to USIs and STJIs. USIs and STJIs have their rows stored on different AMPs than the base table rows and MultiLoad does not support that.

Another consideration when choosing between a NUSI and a Join Index is that for a NUSI, all the columns in a multi-column index have to be specified in the WHERE clause or the optimizer will not use it. For a Join Index, we do not have to specify all the columns for the optimizer to use it.

Base Table and Single Table Join Index with Different Primary Indexes

In this example of STJI usage, the Store Table has Store, Item, DateSold, QtySold, and QtyOnHand columns available in the base table.

STJI				
store	item	date_sold	qty_sold	on_hand_qty
101	123456	date1	0	5
101	123456	date2	0	5
101	123456	date3	0	5
101	123456	date4	1	4
101	123456	date5	0	4
101	123456	date6	0	4
.	.	.		
.	.	.		
.	.	.		
152	111111	date1	0	3
152	111111	date2	0	3
152	111111	date3	0	3
152	111111	date4	0	3
152	111111	date5	0	3
152	111111	date6	0	3
.	.	.		
.	.	.		
.	.	.		

Figure 80: Store Table Columns

We can create the Single Table Join Index with the same columns as the base table, but use a different Primary Index than we used for the base table. The Primary Index of the base table is on Store, Item, and DateSold. The Primary Index for the Join Index is on Store and Item. By creating the STJI with a PI on Store and Item, all the Item and Store combinations, with different sale dates, are grouped together in the same data block for a given Item and Store.

The PI of the STJI is an alternate access path to the data and the data is redistributed one time during the creation of the JI. The huge benefit is that we avoid row redistribution every time the query executes. Essentially, we get to have two copies of the table that we indexed two different ways. Think of the STJI as a second copy of the table, but with a different Primary Index, and that it is maintained automatically.

> **Store table PI = (store, item, date_sold)**
> Rows are redistributed and sorted to get all rows with same store and item and date on same AMP
>
> **STJI PI = (store, item)**
> Rows have been redistributed and grouped together at the time STJI was built

Figure 81: Store Table and STJI Data Distribution

To select certain items sold in a set of stores for a given set of dates, the base table's Primary Index is Store, Item, and DateSold. The rows are sorted and redistributed to get all the rows for the same Store, Item, and Date onto the same AMP.

```
SELECT Item,
COUNT (DISTINCT(StoreNo))
FROM    SalesHistory
WHERE OnHandQty > 0
AND    QtySold > 0
AND    Item      IN (x,y,z)
AND    DateSold IN (a,b,c)
AND    Store    IN (d,e,f)
GROUP BY 1 ORDER BY 1;
```

Figure 82: Find Items Sold in a Set of Stores for a Given Set of Dates

Here is another scenario for the STJI with a Primary Index of Store and Item. Again, the rows were redistributed and grouped once at JI create time. Now when we SELECT an Item and count the stores with their history of QtySold, the query uses the Join Index, and runs ten times faster than if it were accessing the base table.

Same Primary Index for Single Table Join Index – Acts Like a NUSI

Using the same Primary Index for the STJI and the base table causes the JI row to be stored on the same AMP as the base table row. That makes the STJI similar to the NUSI because the STJI is AMP local, as the NUSI is AMP local. This is very useful for partial covering, where we retrieve a portion of the answer set from the STJI table. The STJI is used to qualify some data before the base table is accessed. The JI covers the non-covered columns by going to the base table. This can greatly reduce the number of rows retrieved from the base table. The Join Index acts like a NUSI doing AMP-local processing with no Teradata BYNET® traffic, but in this case, the NUSI would not have been accessed for a partial search.

Having the same PI for a JI and base table is useful for scoring data. For example: we create a list for telemarketing with only 2,000 names from the 20 million in our data warehouse. Rather than doing a Full Table Scan of the base table, we can scan the much narrower STJI table, get our values, and then go to the base table for the rest of the data.

Global Join Index – Using the LIKE Clause on a STJI

Another example of creating a Join Index is to make it look just like the base table. We can call it LikeTab:

```
CREATE JOIN INDEX LIKETAB as
    SELECT   CarLicense, PICol, ROWID
    FROM     CustomerInfo
    PRIMARY INDEX (PICol) ;
```

Figure 83: Create a Join Index that Looks Like the Base Table

Notice that the RowID is included in the Join Index definition, making this a Global Join Index.

The query accesses the CustomerInfo Table and the Primary Indexes are the same for the Global Join Index and the base table.

```
SELECT *
FROM CustomerInfo
WHERE CarLicense LIKE 'ABC%';
```

Figure 84: Query that Causes a Full Table Scan of the Join Index

When we SELECT from the CustomerInfo Table using a LIKE clause, the query will do a Full Table Scan of the very narrow Join Index table. It qualifies the rows with the license plates that begin with ABC, and uses the RowID to join back to the CustomerInfo Table. The rows are already on the same AMP because the JI and CustomerInfo Table have the same Primary Index. The optimizer does not use a NUSI to scan when it has the LIKE condition; it chooses to scan the base table instead. Scanning the Join Index gives the query much faster access and improves performance.

Build a STJI with a Column for LIKE Plus RowID of Base Table

One real-life example is a customer who wanted OLTP-type response times: seconds, not minutes. The LIKE clause on the base table caused a Full Table Scan, so with 40 million rows on a 19 column table on a two node system it took one minute to execute. We built a Global JI that included the column for the LIKE scan and the RowID of the base table rows. The LIKE clause accessed the narrower JI table, and that gave a 4-second response time. That is a 56-second savings per query. Multiply that times every time the query is executed, and that is a considerable performance gain.

Value Ordered Indexes for Range Searches

Value Ordered NUSIs (VONUSIs) are very efficient for range searches with inequality conditions on the Secondary Index column set.

The value-ordering option is available on both NUSIs and STJIs with a numeric restriction of 4 bytes for Integer values only. No character data is allowed and there is a limit of 64 columns.

```
CREATE INDEX OrdDate (OrderDate)
ORDER BY VALUES      (OrderDate)
ON ORDERS
;
```

Figure 85: Syntax for a Value Ordered Index

The VONUSI and the Value Ordered Single Table Join Index (VOSTJI) provide great access to data when selecting a range of values.

For example, the Invoice Table has 60 million rows which represent 1500 days times 100 outlets times 400 sales per day. The Invoice_Item Table has 4 items per sale or 60 Million times 4 which are 240 million rows.

In this case, the NUSIs are not joined, they are used for accessing the data, but the base tables are joined. The Join Indexes, however, can be used without access to base tables, which is much more efficient.

cdate	storeno	saleno	invoice_id	itemno	quantity	price
103090I						
103090I						
103090I						
103090I						
103090I						
103090I						
103090I						
103090I						
103090I						

Figure 86: Table Data Rows are Hashed Query Does Full Table Scan

Table data rows are hashed on the Primary Index. To satisfy the query, the system does a Full Table Scan. It looks at all the rows selecting the ones that qualify for the specified date.

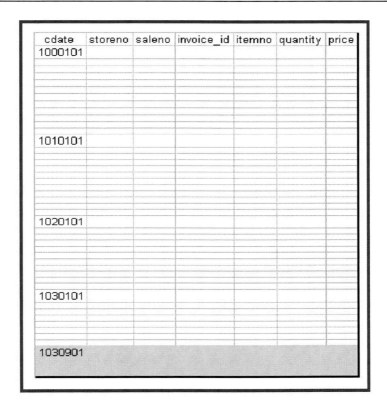

Figure 87: Value Ordered NUSI/STJI – Query Scans Only Specified Portion of Table Based on Specified Value

For a Value Ordered NUSI or a Value Ordered STJI, all the rows for a specific date are grouped and stored together. Access to the data is much faster.

SPARSE Join Indexes

Sparse Join Indexes are used to limit the rows in the JI table. Sometimes it makes sense to index only the rows that are used most frequently; for example, indexing only the most recent quarter for time-based queries. The Sparse Join Index feature eliminates some rows, for example NULLS, making access even faster. A Sparse Join Index has fewer rows, and is smaller. It takes less work to maintain, and updates are quicker since there are fewer rows to update. The benefit is faster access, reduced storage requirements for the Join Index, and reduced maintenance costs for updates.

To make the Join Index sparse, use a WHERE clause predicate at create time. This allows us to index only the rows we want for the portion of the table that is used most frequently, or the rows that satisfy a suite of queries. The following example extracts only information for the year 2011.

```
CREATE JOIN INDEX SparseJI AS
SELECT StoreId
       , ProdId
FROM    Sale
WHERE EXTRACT (Year FROM SaleDate) = 2011;
```

Figure 88: Create a Sparse Join Index

Additionally, Join Indexes may include aggregates such as TotalSalesAmt for products in each store for the year 2011.

```
CREATE JOIN INDEX AggSparseJI AS
  SELECT StoreId, ProdId, SUM(SalesAmt) AS TotalSalesAmt
  FROM    Sale
  WHERE EXTRACT(Year FROM SalesDate) = 2011
  GROUP BY StoreId, ProdId;
```

Figure 89: Create an Aggregated Sparse Join Index

Sparse Join Indexes support high frequency queries requiring short response times.

Sparse Indexes reduce the storage requirements for a Join Index and also the maintenance costs for updates because there is a relatively small portion of the table being stored. Sparse Indexes make access faster since the size of the Join Index is significantly smaller than a full Join Index, especially if the base table is very large.

A Sparse Index can focus on the portion of the table(s) that are most frequently used.

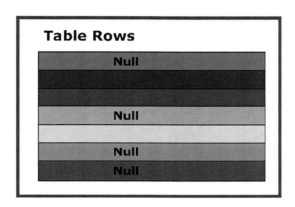

Figure 90: Table Row Size

This depicts the base table, including the column that is to be indexed. The table is fairly wide and contains all the data rows.

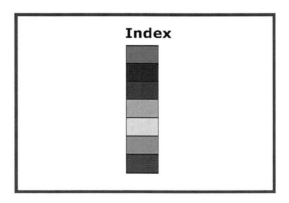

Figure 91: Index Size

If we add a Join Index on the table, the JI table is narrower than the base table because it only has one column. The JI table still contains all the rows, but is much smaller than the base table.

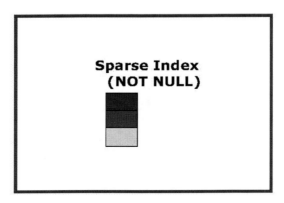

Figure 92: Sparse Index Size

Here we eliminated all the rows that are NULL, making the Sparse Join Index even smaller than the JI. Consider an even smaller table by creating a Sparse Join Index for a specific date. Perhaps we only want to look at yesterday's sales. We can specify a WHERE clause with yesterday's date. That way, we when want information about yesterday's sales, we only have to look at yesterday's data. The optimizer uses the Sparse Join Index for access and that is much faster than either the base table scan or the STJI scan, because it looks at a much smaller amount of data.

Sparse Single Table Join Index – A Form of Horizontal Partitioning

We can build a Sparse Single Table Join Index as a form of horizontal partitioning to enable us to qualify rows in a Join Index.

```
CREATE JOIN INDEX SJIActiveSales
AS
        SELECT *
        FROM   Sales
        WHERE Status = 'ACTIVE';
```

Figure 93: Sparse Join Index Syntax

The example we have is from a Sales Table.

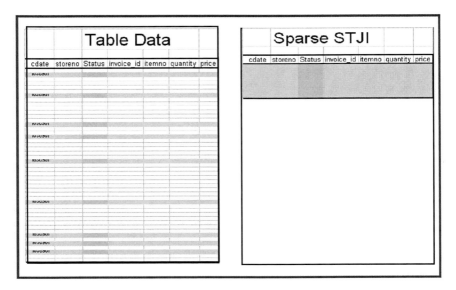

Figure 94: Sparse Single Table Join Index, a Form of Horizontal Partitioning

The left side of Figure 94 shows a populated table with Status column highlighted for all active accounts. We create a Join Index, JIActiveSales, by selecting all the columns from the table where the status is 'active'.

The status column is populated with various values such as 'active', 'closed', 'pending', etc. The Sparse Join Index has just the qualifying rows, where the status is 'active'. In the previous "Sparse Index for Not Null" example, we had all the table rows, but the qualifying rows were all grouped together. In this example with the Sparse Join Index, the Join Index only contains the rows for active sales.

Using a communications industry example, there are residential customers who represent 94% of our customers who make 50% of all the phone calls.

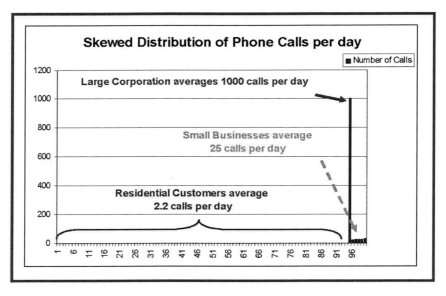

Figure 95: Skewed Data Distribution

Assuming the application requires a quick response time for residential customer calls, we can use a Sparse Single Table Join Index instead of a NUSI. The Sparse STJI saves on space and maintenance costs. It contains the residential customers who on average make fewer than 2.5 calls per day. If we use the Join Index for the large corporations that account for a greater number of calls, the Sparse STJI would not be very beneficial. However, the Sparse STJI might be beneficial for small business customers who average about 25 calls per day.

Sparse Single Table Join Index – Three More Examples

A Sparse Single Table Join index may be useful for an insurance company application that stores 10 years of history. Clients renew their coverage every six months. A Join Index on the current paid policies represents 1/20th of the customer policies, and only 1/20th of the data goes into the Sparse Single Table Join Index.

In parts manufacturing, not every part is available in inventory. Generally speaking, fewer than 1% of parts are available at a given time. We could consider a Sparse Join Index on just the available parts, because usually that is the information that is needed most often, rather than the parts that might be out of stock. Retrievals from the Part Table would look for *available* parts and the Sparse STJI would be a great benefit.

In the Retail environment, the concern is filled orders versus the unfilled Orders. Usually less than 0.5% of orders are unfilled. The query would be looking for just the unfilled orders and again, a Sparse Index would be very beneficial.

Value Ordered Sparse Single Table Join Index

To illustrate a point, we've gotten a little extreme with the next Transportation example. We have both vertical and horizontal partitioning at the same time. We have created a Sparse Join

Index that qualifies only the active rows and the Join Index only has the columns that we need to see.

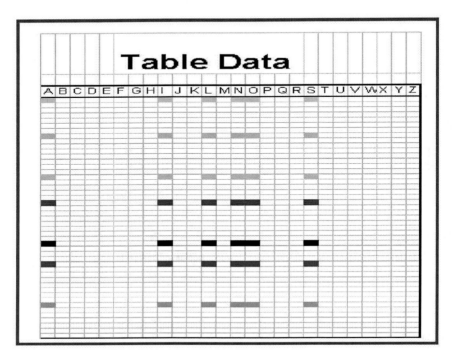

Figure 96: 12-Month History of All Flyers and All Flights

The base table has data in it that shows 12 months of history for all the fliers and all the flights they have taken.

Figure 97: Today's Flyers and Today's Flights

This Value Ordered Sparse Single Table Join Index shows only today's flights with today's fliers.

Value Ordered Sparse STJI on PPI Table

Another interesting scenario is a Value Ordered Sparse Single Table Join Index on the partition of a Partitioned Primary Index Table.

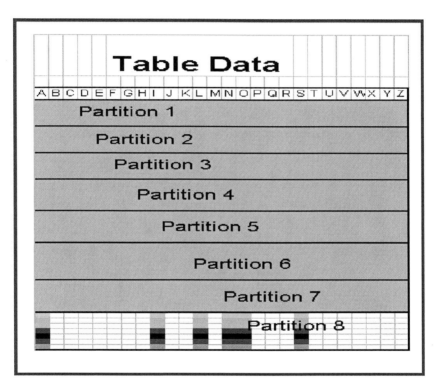

Figure 98: Daily partition with 100M Rows

A Retailer adds 100M rows/day to a partitioned table, and only wants to access a subset of the latest data. Accessing this Partition Primary Index eliminates seven of the eight partitions.

Figure 99: 100M Row Subset of the Latest Data

If we add a Sparse Value Ordered STJI on the PPI Table, it gives us data from today and only the columns we want to see. This results in not only good partition elimination, but a very small Join Index to access.

Creating a Compressed Sparse Multi-Table Join Index

Let's build another Sparse Join Index. In this example, we create a Compressed Sparse Join Index on a specific portion of the table.

```
CREATE JOIN INDEX    Cust_Ord_SJI AS
    SELECT           (c_CustId, c_LName),
                     (o_OrderId, o_OrderStatus, o_OrderDate)
    FROM             Customer C
    INNER JOIN       Orders O
    ON               c_CustId = o_CustId
    WHERE            EXTRACT (YEAR FROM o_OrderDate) =
                     EXTRACT (YEAR FROM CurrentDate)
    PRIMARY INDEX    (c_CustId);
```

Figure 100: Create Sparse Join Index for the year 2011

Notice in the WHERE clause, we EXTRACT the YEAR FROM the OrderDate where the year is the Current Date. If the year is 2011, the Join Index only has orders for 2011.

```
SELECT      c_CustId
            , c_LName
            , o_OrderDate
FROM        Customer C
INNER JOIN  Orders O
ON          c_CustId = o_CustId
WHERE       o_OrderDate = '2011-01-27'
AND         o_OrderStatus = 'O';
```

The join index will be used for this SQL and the EXPLAIN estimated cost is 0.06 seconds.

Figure 101: Query for 2011 Open Orders

To execute a query selecting open orders in 2011, we use the Join Index.

```
SELECT      c_CustId
            , c_LName
            , o_OrderDate
FROM        Customer C
INNER JOIN  Orders O
ON          c_CustId = o_CustId
WHERE       o_OrderDate = '2010-12-18'
AND         o_OrderStatus = 'O';
```

The tables will have to be joined for this SQL and the EXPLAIN estimated cost is 0.44 seconds.

Figure 102: Query for 2010 Orders

What if we run the query with an order from 2010? That query completes, and the user gets the result, but the query accesses the base table instead of the Join Index, because there are no rows for 2010 in the Sparse Join Index.

Creating a Sparse Join Index on a Partitioned Table

We have a great deal of flexibility with Teradata Database Indexes. We can put a Sparse Join Index on a partitioned table and on a Sparse Join Index. The Sparse Join Index can be created for a partition or partitions, and only the "partitions of interest" are scanned.

Creation time for the Sparse Join Index on a partition is fast because of partition elimination.

This example shows partitioning on the OrderPPI base table.

```
CREATE SET TABLE Order_PPI
  ( o_OrderId       INTEGER NOT NULL,
    o_CustId        INTEGER NOT NULL,
         :                :
    o_OrderDate     DATE FORMAT 'YYYY-MM-DD' NOT NULL,
         :                :
    o_comment       VARCHAR(79))
PRIMARY INDEX     (o_OrderId)
PARTITION BY RANGE_N
  ( o_OrderDate BETWEEN DATE '2002-01-01'
    AND DATE '2011-12-31' EACH INTERVAL '1' MONTH ) ;
```

Figure 103: OrderPPI Table Partitioned By Month

The OrderPPI Table has 10 years of data that is partitioned by month.

```
CREATE JOIN INDEX Order_PPI_JI AS
    SELECT           o_OrderId
                   , o_CustId
                   , o_OrderStatus
                   , o_totalprice
                   , o_OrderDate
    FROM             Order_PPI
    WHERE            o_OrderDate
                     BETWEEN '2011-01-01'
                     AND '2011-03-31'
    PRIMARY INDEX  (o_CustId);
```

Figure 104: Sparse Join Index on OrderPPI Table for Q1 2011

When we create this Sparse Join Index, it is not partitioned, but it is built for a specific Order where we want to access the first three months of 2011. One of the benefits of building this Sparse

Join Index on the partitioned table is that the creation of the Join Index is very fast.

```
7) We do an all-AMPs RETRIEVE step from 3 partitions of MyDB.OrderPPI with a
   condition of ("(MyDB.OrderPPI.o_OrderDate <= DATE '2011-03-31') AND
   (MyDB.OrderPPI.o_OrderDate >= DATE '2011-01-01')") into Spool 1 (all_amps), ...

                3 partitions are scanned to build the JI.
```

Figure 105: EXPLAIN Output Showing Partition Elimination

The EXPLAIN of the JI CREATE shows we only have to look at three of the 120 partitions in the OrderPPI base table.

Creating a Sparse Join Index with Partitioning

We also have the option to partition a Sparse Join Index.

```
CREATE JOIN INDEX OrderPPISJI2 AS
    SELECT        o_OrderId
                , o_CustId
                , o_OrderStatus
                , o_totalprice
                , o_OrderDate
                , o_ClerkId
    FROM          OrderPPI
    WHERE         o_OrderDate BETWEEN '2011-01-01'
                  AND '2011-12-31'
    PRIMARY INDEX (o_CustId)
    PARTITION BY RANGE_N
       (o_OrderDate BETWEEN
        DATE '2011-01-01' AND DATE '2011-12-31'
        EACH INTERVAL '1' DAY) ;
```

Figure 106: Partitioned Sparse Join Index Syntax

Above, we create a Sparse Partitioned Join Index where we select columns from the OrderPPI Table for the year 2011. We partition this Join Index with daily intervals, giving us 365 partitions. The

Sparse Join Index has a different partitioning expression than the base table. A query for day-to-day comparisons uses this Join Index.

```
CREATE JOIN INDEX OrderPPISJI2 AS
    SELECT          o_OrderId
                    , o_CustId
                    , o_OrderStatus
                    , o_totalprice
                    , o_OrderDate
                    ,o_ClerkId
    FROM            Order_PPI
    WHERE           o_OrderDate BETWEEN '2011-01-01'
                    AND '2011-12-31'
    PRIMARY INDEX   (o_CustId)
    PARTITION BY RANGE_N
        (o_OrderDate BETWEEN DATE '2011-01-01'
                    AND DATE '2011-12-31'
                    EACH INTERVAL '1' DAY )
    INDEX (o_ClerkId);
```

Figure 107: Partitioned Sparse Join Index with a NUSI Syntax

We can also place NUSI on a Sparse Partitioned Join Index. The query could be qualified using the NUSI on the Sparse Partitioned Join Index for specific orders, or sales for a particular clerk.

Global Join Index on Multiple Tables

The term Global Index refers to a Join Index which includes the keyword RowID as a data column within the Join Index.

Global Join Indexes can be used in two ways. Firstly, they can join back to the base table using the RowID. Secondly, they can serve as an alternative to a Secondary Index. Sometimes this form of Global Join Index is referred to as a Hashed NUSI.

```
CREATE SET TABLE Customer
   ( c_CustId         INTEGER NOT NULL
   , c_LName          VARCHAR(15)
   , c_fname          VARCHAR(10)
   , c_address        VARCHAR(50)
   , c_city           VARCHAR(20)
   , c_state          CHAR(2)
   , c_zipcode        INTEGER)
UNIQUE PRIMARY INDEX ( c_CustId );
```

Figure 108: Create Customer Table Syntax

```
CREATE SET TABLE Orders
   ( o_OrderId        INTEGER NOT NULL
   , o_CustId         INTEGER NOT NULL
   , o_OrderStatus          CHAR(1)
   , o_totalprice     DECIMAL(9,2) NOT NULL
   , o_OrderDate      DATE
                      FORMAT 'YYYY-MM-DD' NOT NULL
   , o_OrderPriority SMALLINT
   , o_Clerk          CHAR(16)
   , o_ShipPriority SMALLINT
   , o_Comment        VARCHAR(79))
UNIQUE PRIMARY INDEX ( o_OrderId );
```

Figure 109: Create Orders Table Syntax

In the first example of a Global Join Index, we use the two tables, Customer and Orders.

```
CREATE JOIN INDEX  CustOrdGJI AS
    SELECT          (c_CustId, c_LName, C.ROWID AS crid),
                    (o_OrderId, o_OrderStatus, o_OrderDate, O.ROWID AS OrId)
    FROM            Customer C
    INNER JOIN      Orders O
    ON              c_CustId = o_CustId
    PRIMARY INDEX   (c_CustId);
```

Figure 110: Create Compressed Global Join Index Syntax

In this example we create the Compressed Global Join Index with CustId and LName, the columns most frequently used in queries. This basically classifies the Global JI as a Covered Join Index. Also notice we include the RowID in the fixed portion of the CREATE statement.

The fixed portions of the rows increase by 10 bytes, because RowIDs in Join Indexes are always 10 bytes long, even if the base table is non-partitioned. We also include the RowID in the repeating portion for each of the Orders. This puts information in the Join Index that is used to join back to the base table.

Global Index as a "Hashed NUSI"

This example of a Global Join Index is as a Hashed Non-Unique Secondary Index. Our Orders Table typically has two orders per customer. We build a NUSI on CustId on the Orders Table.

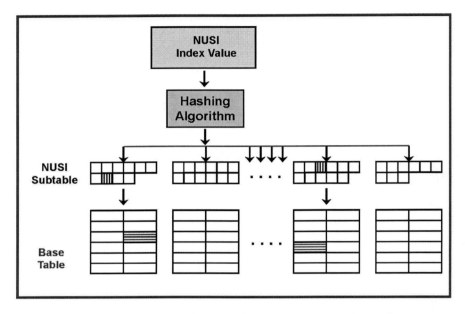

Figure 111: All AMP Operation Avoids a Full Table Scan

The NUSI access causes an all AMP operation. Each AMP looks at its subtable. If we only have two orders (highlighted in the base table above), then only two of the AMPs do productive work. If we had 2,000 AMPs, 1998 of them would do an I/O on the NUSI subtable, but find no rows. It might be more efficient to create a Global Join Index, (Hashed NUSI) on this column.

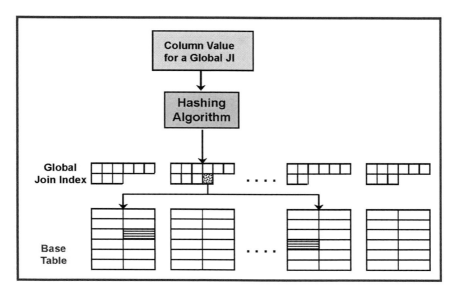

*Figure 112: Global Join Index ("Hashed NUSI")
Uses a Group AMP Operation*

We show a Global JI row created with multiple RowIDs for the base table rows, this example has two RowIDs. When we use this Global JI, three AMPs are involved. We go to one AMP to find the Global Join Index that has the multiple RowIDs. We then use the RowIDs to access two other AMPs in the system to look at base table rows. Three AMPs are needed to satisfy the query, and there are only three physical I/Os.

When does a NUSI become more efficient than this type of Global Join Index?

When the typical rows per value are greater than 50% of the number of AMPs on the system, a NUSI gives us an advantage. If rows per value are less than 50% of the number of AMPs, then the Global Join Index (Hashed NUSI) is more efficient.

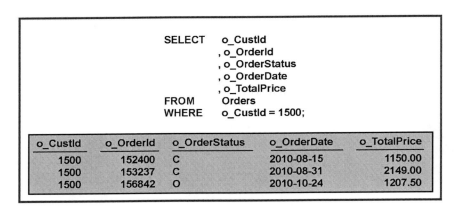

Figure 113: List the Orders for CustId 1500

This EXPLAIN plan shows that the Global Join Index works!

```
  .
  .
  .
3) We do a single-AMP RETRIEVE step from ATLCLH.OrdersGI by way of
   the primary index "ATLCLH.OrdersGI.o_CustId = 1500" with no residual
   conditions into Spool 2 (group_amps), which is redistributed by hash
   code to all AMPs. Then we do a SORT to order Spool 2 by the sort key
   in spool field1. The size of Spool 2 is estimated with high confidence to
   be 14 rows. The estimated time for this step is 0.00 seconds.
4) We do a group-AMPs JOIN step from Spool 2 (Last Use) by way of an
   all-rows scan, which is joined to ATLCLH.Orders. Spool 2 and
   ATLCLH.Orders are joined using a row id join, with a join condition of
   ("Field_1 = ATLCLH.Orders.RowID"). The result goes into Spool 1
   (group_amps), which is built locally on the AMPs. The size of Spool 1
   is estimated with index join confidence to be 4 rows. The estimated
   time for this step is 0.08 seconds.
  .
  .
  .
```

Figure 114: EXPLAIN Showing Use of a Global Join Index as a Hashed NUSI (Partial Listing)

We search the Orders Table for CustId 1500. Instead of using the NUSI the EXPLAIN plan uses the Join Index and retrieves the CustIDs with a value of 1500 and places the RowIDs into Spool 2.

We then use the RowIDs from Spool 2 to join to the Orders Table and return the data rows.

The benefits are row hash locks, group AMP operations, and a RowID join.

Repeating Row IDs in Global Join Index

A Global Join Index (Hash NUSI) is often created in a compressed format. Below we show CustId in the fixed portion and RowID in the repeating portion.

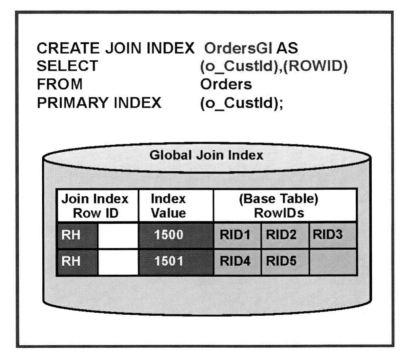

Figure 115: Compressed Global Join Index

We also create this as a Non-Compressed Global Join Index along with other columns that we want partitioning on.

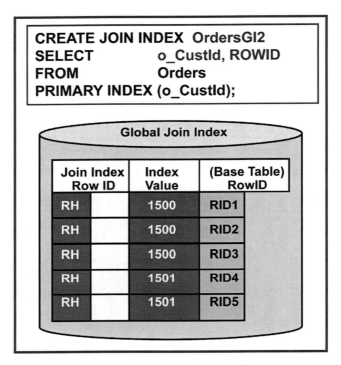

Figure 116: Non-Compressed Global Join Index

If we are not going to partition this Index, we use the compressed format and save a lot of storage space.

```
SELECT    TableName,
          SUM(CurrentPerm) AS SumPerm
FROM      DBC.TableSize
WHERE     DatabaseName = USER
GROUP BY 1
ORDER BY 1;
```

Figure 117: Determine Space Usage for the Global Join Indexes

TableName	SumPerm	
OrderGI	1,065,984	(repeating Row IDs)
OrderGI2	2,019,840	(no repeating Row IDs)

Figure 118: Space Usage Results

Above we see the Permanent Space savings for the compressed Global Join Index vs. the non-compressed Global Join Index.

As with all indexes, the optimizer evaluates the available Join Indexes, and chooses one if it is appropriate for the specific query.

Temporal Support – Move Date in Join Index

Teradata Release 13.10 debuts the database Temporal feature. Teradata Database allows refresh of the content of a Join Index that is defined with CURRENT_DATE or CURRENT_TIMESTAMP.

In a Partitioned Primary Index or Partitioned Join Index, the CURRENT partition may become very large as we add data. Using Temporal, we can realign the partitions to reflect more accurate current data. We can alter partitions containing data using an ALTER TABLE <YourPPIorJoinIndexName> TO CURRENT command to bring the contents up to date. There is no need to drop and recreate the Join Index or the PPI table.

When CURRENT_DATE or CURRENT_TIMESTAMP defines an upper partition boundary, the CURRENT partition may not have all the current data. This minimizes the partition maintenance, and current data is added when the PPI or Join Index is ALTERed to CURRENT.

When CURRENT_DATE or CURRENT_TIMESTAMP defines a lower boundary, the partition may contain old data that is moved out of the current partition when the PPI or Join Index is ALTERed to CURRENT.

This Temporal feature reduces system overhead and maintenance costs.

```
CREATE JOIN INDEX JISales7Days AS
    SELECT  ProdId
            , StoreId
            , SaleDate
    FROM    Sales
    WHERE   SaleDate > CURRENT_DATE – INTERVAL '7' DAYS
    AND     SaleDate <= CURRENT_DATE;
```

Figure 119: Create Join Index with CURRENT_DATE Syntax

For example, create a Join Index based on 7 days of sales for September 1-7.

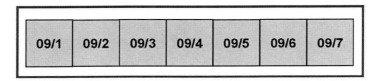

Figure 120: Join Index with Seven Days of Data

Each of the seven days of data has its own partition in the Join Index. What happens if we ALTER to CURRENT?

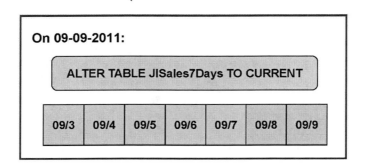

Figure 121: ALTER Join Index to CURRENT

If the JI is altered on September 9th, the two oldest dates are moved out and two new, current dates, are included in the current partition.

Join Index Type Review

We have covered a lot of ground on Join Indexes. Let's do a quick recap.

Simple Join Indexes are automatically updated when the base table rows are updated. The optimizer evaluates and chooses whether or not to use a Join Index.

Single Table Join Indexes are built on a single table. They are used primarily for covering queries (base table RowIDs are optional) and can be hashed on a user-defined Primary Index.

Multi-Table Join Indexes can store covering data from as many as 64 base tables. NUSIs can be defined on the Multi-Table Join Indexes, and the user can define the PI column.

Aggregate Join Indexes may include SUM and COUNT values (from which Averages may be calculated) on one or more of its columns.

Aggregate Join Indexes may be defined on:

- A single table which is a columnar subset of a base table, with aggregates automatically maintained by the software.

- Multiple Tables which are a columnar subset of as many as 64 base tables with aggregate columns automatically maintained.

- Sparse Join Indexes which are defined with a WHERE clause that limits the number of base table rows included in the JI.

Chapter Seven
HASH INDEXES

Hash Indexes are not very widely utilized. They are similar in their usage to Single Table Join Indexes, but aggregation, secondary indexing, and triggers are not permitted on Hash Indexes, so most environments use STJIs instead. One unique thing about the Hash Index, is that it automatically includes the base table's Primary Index value. The Teradata Database optimizer uses this information to hash and join back to the base table or to another table. Hash Indexes allow for the creation of a full or partial replication of the base table, with a Primary Index specified at the time the Hash Index is created. Hash Indexes are useful for joins and index covering.

Consider Hash Indexes for improving query performance. The Hash Index does provide a space-efficient index structure and can be hash distributed to AMPs in a number of ways. The Primary Index of a Hash Index can be the same or different than the Primary Index of the base table.

Having the same PI is useful for joining to the base table. Having a different PI, is useful for joining to other tables. When the PI of the Hash Index is the same as the PI of the table it will be joined to, the data is redistributed once when the Hash Index is built. No further redistribution is needed for queries.

Hash Indexes are most useful when the number of rows for a given value in the Hash Index is less than half the number of AMPs on the system. That requires knowledge of the system and data demographics.

Teradata MultiLoad and FastLoad utilities do not support target tables with Hash Indexes. To work around this, FastLoad data into an empty table, and then use either an INSERT/SELECT or MERGE to update the populated target table that has the Hash Index.

Hash Indexes support value ordering.

Below (Figure 122), is the syntax to create a Hash Index on the Color column of the Car Table.

```
CREATE HASH INDEX CarColorInfoHI
       (Color) ON Car BY (Color)
ORDER BY (Color);
```

Figure 122: Create Hash Index Syntax

Figure 123 show the ORDER BY options for Hash Indexes.

```
ORDER BY                ColumnName;
ORDER BY    HASH     (ColumnName);
ORDER BY    VALUES   (ColumnName);
```

Figure 123: Hash Index ORDER BY Options

There are three ordering options for Hash Indexes. ORDER BY ColumnName defaults to ordering by the hash of the column value. ORDER BY HASH (ColumnName) stores the rows in column value hash sequence and is good for range searches. ORDER BY VALUES (ColumnName) sorts on the column value and is good for data value searches.

Hash Index Example

A Hash Index is useful for a Marketing Campaign where it can be used to reduce a list of prospects. We execute successive qualifying queries (against the Hash Index) until the desired number of prospects is achieved and then we retrieve detailed data from the base table.

Hash Indexes are similar to secondary indexes in that they are created for a single table only. The CREATE syntax is very similar to a secondary index, and Hash Indexes may cover a query without accessing the base table rows.

Hash Indexes are similar to Join Indexes, because they redistribute joinable rows to a common location. When we choose the PI of the Hash Index, we determine the distribution and sequencing of the rows, making Hash Indexes very similar to Single Table Join Indexes.

Hash Indexes can be very useful, however, we may decide that one of the many forms of Join Indexing afford us all the flexibility we need in getting the best access to our data.

Chapter Eight
NO PRIMARY INDEX TABLES

Teradata Release 13 offers the option of creating tables without Primary Indexes. This type of table is useful for staging data. When it is difficult to choose a PI because it is only adequate for a subset of the transformation operations, having no Primary Indexes defined on tables simplifies ELT operations. The data can be loaded quickly, because the Teradata Database does not have to hash the Primary Index value for each row in the table.

If there is no natural PI choice for a staging table that is used for a large number of ELT processes, skewing will result when the data is loaded. A NoPI table may allow the data to load more evenly.

NoPI Table Load

Tables are distributed based on the hash of their Primary Index. Since NoPI tables have no Primary Index defined, the data is distributed across the AMPs by a specialized process added in Teradata Release 13.

FastLoad, for initial NoPI table population, runs at the AMP level, processing data a block at a time. Data blocks for NoPI tables are distributed randomly across the AMPs using the already mentioned specialized process running on the AMPs. FastLoad uses two phases. In Phase 1, it distributes the data and writes it to disk. In Phase 2, it sorts the data. For a NoPI table, FastLoad eliminates Phase 2 completely, making FastLoad much faster than row level processing.

When using Teradata TPump and SQL to load or update data, processing is done a row at a time by the PE, using random generator code. As rows are inserted into a populated NoPI table, they append to the end of the table. Rows are not organized or sorted based on RowHash.

CREATE NoPI TABLE with New Syntax

The NoPI CREATE TABLE syntax in Figure 124 specifies the table name, a list of columns, and includes the NO PRIMARY INDEX clause.

```
CREATE TABLE <table_name> (<column1>   <column1_datatype>,
                           <column2>   <column2_datatype>,
                           ... )
NO PRIMARY INDEX;
```

Figure 124: Create NoPI Table Syntax

NoPI tables are automatically created as MULTISET tables and if we specify the table as a Set Table, an error is generated. Internally, the TableKind is set to type 'O' for NoPI tables, instead of type 'T'.

If the NoPI table is created with a Primary Key or has unique constraints, the Teradata Database creates USIs for the Primary Key or the constraints.

The CustomerTemp NoPI Table in Figure 125 is a staging table. Fallback is available, all of the columns are defined, and NO PRIMARY INDEX is specified.

```
CREATE TABLE CustomerTemp, FALLBACK
       (Custid         DECIMAL(18,0)
       ,LastName       VARCHAR(25)
       ,FirstName      VARCHAR(15)
          :
       ,CountryCode    CHAR(3)
       ,PostalCode     CHAR(10))
NO PRIMARY INDEX;
```

Figure 125: Create CustomerTemp NoPI Table

In the example in Figure 126, we create the OrderTemp NoPI Table. It is a copy of the Order Table which is another table found in the database.

> **CREATE TABLE OrderTemp AS Orders**
> **WITH DATA *NO PRIMARY INDEX*;**

Figure 126: Create OrderTemp NoPI Table

The new OrderTemp Table does not have a PI. If the original Orders Table is evenly distributed, then OrderTemp Table is also evenly distributed. Creating a table or making a copy of a table, is a block-level copy operation, performed within the AMPs. The data remains on the same AMP where it was originally stored for the Orders Table. If the Orders Table is skewed, then OrderTemp is also skewed.

The Row ID for a NoPI Table

As with other tables, the RowIDs in a NoPI table are eight bytes in length. *Figure 128* shows the first 20 bits represent the Hash Bucket number which is the same for all of the rows of the NoPI table on a specific AMP. The last 48 bits of the RowId represent the uniqueness value. We can see that each time a row is stored on an AMP, the uniqueness value increases by one, and the row is appended to the end of the table. The row for customer Gary Garcia was added before the row for Kara Chrystal. Both have the same Hash Bucket number as the other rows in this NoPI table. Each time a row is added, the Uniqueness value is incremented by one. Row distribution to the AMPs is based on the hash of the DBQL QueryId.

Teradata® Database Index Essentials

Row ID for NoPI table	Hash Bucket 20 (or 16) bits	Uniqueness Value 44 (or 48) bits			
Each row still has a Row ID as a prefix.	Row ID		Row Data		
	Hash Bucket	Uniqueness	CustNo	LastName	FirstName
	000E7	00000000001	001018	Reynolds	Tiffany
	000E7	00000000002	001020	London	Evan
Rows are logically	000E7	00000000003	001031	Vazquez	Joe
maintained in Row ID	000E7	00000000004	001014	Jacobs	Ryan
sequence.	000E7	00000000005	001012	Garcia	Gary
	000E7	00000000006	001021	Chrystal	Kara
	:	:	:	:	:

Figure 127: Row ID for a NoPI Table

AMP-Level View of Multiple NoPI Tables

The AMPs in *Figure 128* are numbered 0-26, and there are two NoPI tables. The TableId of the first table is 00089A, and the second table is 00089B. The hash value for each row on AMP3, regardless of the table, starts with 000E7.

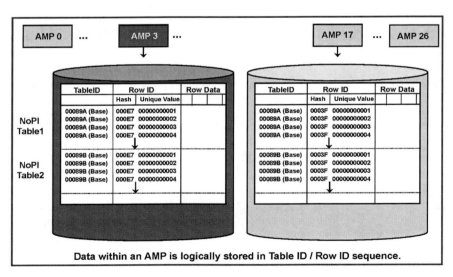

Figure 128: Multiple NoPI Tables at the AMP Level

Looking at AMP17, the hash is 0003F for the same two NoPI tables.

Populating a NoPI Table from Existing Tables

Scenario 1: When populating the OrderTemp NoPI Table, using a simple INSERT/SELECT from a single existing Orders Table, the data is copied at the block level within the AMP. If the existing table is skewed, the target NoPI table is skewed in the same way.

Scenario 2: When populating the OrderTemp NoPI Table, using INSERT/SELECT from multiple tables (via a join), the data from the source tables are joined together in spool. The blocks of spool are then copied into the NoPI table on each AMP. If the spool is skewed by the join, then the target NoPI table is skewed in the same way.

In summary, an INSERT/SELECT into a NoPI table does not redistribute the data between the AMPs. The NoPI random generator code is not used in these scenarios. The data is stored on the same AMP as the table it was copied from.

Populating a PI Table from an Existing NoPI Table

Scenario 1: OrderTemp, a NoPI table, is populated via FastLoad. FastLoad randomly distributes blocks of data between the AMPs and the data is evenly distributed.

When populating Orders, a target PI table, the rows are redistributed on the AMPs based the hash of the PI values. If the Orders target table has a Unique Primary Index, then the target table is distributed evenly across the AMPs. If the target table has a NUPI, then the distribution is dependent on the uniqueness of the NUPI.

Scenario 2: OrderTemp is a NoPI table and it is populated using an INSERT/SELECT from an existing table with a NUPI on OrderId. The NoPI table rows are stored on the same AMPs as they are stored in the source table with a NUPI on OrderId.

If we then populate another table with a NUPI on CustId, from the OrderTemp NoPI table, the rows are redistributed on the AMPs based on the hash of the CustId PI values. As always, the evenness of distribution is dependent upon the uniqueness of the NUPI values.

If we populate yet another table with OrderId as its NUPI from the NoPI OrderTemp Table, which already has the same NUPI as the original source table, row redistribution still occurs. The rows remain on the same AMP, however, a RowID is generated based on the PI value, and the rows are sorted into RowID sequence. Therefore, this is not a block copy operation.

In summary, an INSERT/SELECT into a PI table from a NoPI table requires data redistribution.

SQL Commands and NoPI Tables

Following are a few examples of SQL command options for the OrderTemp NoPI table.

OrderTemp (no primary index)
 	
CREATE INDEX (OrderDate) ORDER BY VALUES ON OrderTemp;
    Secondary indexes are created/dropped just like a PI table.
 	
ALTER TABLE OrderTemp, FALLBACK;
ALTER TABLE OrderTemp, NO FALLBACK;
    FALLBACK can added or dropped just like a PI table.
 	
COLLECT STATISTICS ON OrderTemp COLUMN (OrderDate);
    Statistics can collected or dropped just like a PI table.
 	
SELECT * FROM OrderTemp WHERE OrderId = 102221;
    Assuming a secondary index has NOT been created on OrderId, this query will result in a full table scan.

Figure 129: SQL Commands and NoPI Tables

We can create NUSIs on NoPI tables, although they are not used frequently. We can also create NUSIs with ORDER BY VALUES, otherwise known as Value Ordered NUSIs.

If we SELECT from a NoPI table that has NUSIs, the optimizer makes its access choice for the NoPI table as it would for a table with a defined Primary Index. If there are no NUSIs on the NoPI table, Teradata Database uses a Full Table Scan.

Statistics can be collected on NoPI tables and FALLBACK protection is supported.

Chapter Nine
INDEX REVIEW

Unique Primary Indexes are very efficient; they access one row from one AMP and require no spool files.

Non-Unique Primary Indexes are efficient if the number of rows per value is reasonable and there are no severe spikes in the data distribution. PIs, unique or non-unique, are always a one-AMP operation. NUPIs can return multiple rows. A spool file is created if needed.

No Primary Index Tables are temporary staging tables for loading data. Access is via a Full Table Scan or NUSIs.

Unique Secondary Indexes are very efficient because they return one row, no spool file is required, and data is accessed using two AMPs.

Non-Unique Secondary Indexes are only efficient if the rows accessed are a small percentage of the total data rows in the table. NUSI access is an all-AMP operation that brings back multiple rows, and if needed, creates a spool file. Statistics are needed for the optimizer to use NUSIs.

Partitioned Primary Indexes can be Single or Multi-Level with up to 65,535 partitions spanning 15 levels of depth. Teradata Release 14 supports 9.2 quintillion partitions.

Partition Scans are very efficient because of partition elimination. Access is on all AMPs and all rows, but only for selected partitions.

Join Indexes can be on Single or Multiple tables. They can be Value Ordered for range searches. A WHERE clause makes it a Sparse JI, reducing the number of rows stored in the index. Adding a RowID makes it a Global Join Index and is useful for joining back to a base table.

Full-Table Scans take place if there is no or insufficient information in the WHERE clause, there are no STATISTICS collected, or the optimizer chooses not to use an index. Full Table Scans touch all rows on all AMPs and are very efficient since each row is touched only one time. The spool file may equal the size of the table.

Optimizing Access Paths. The optimizer creates a plan and chooses the fastest access method possible to access the data. It may use indexes or a Full Table Scan. Collecting Statistics helps the optimizer to make the best decisions.

FAQs

Q – *We want a PI that is used frequently in joins, but if it is not unique, how much skewing is acceptable to avoid creating an identity column?*

A – The size of the table and your system configuration determines if skewing is a factor to consider. A large table with skewing that exceeds approximately 10-20% of the total table data, where some of the AMPs have 20% more data than other AMPs, could be considered a skewing issue. That is because there is more data and more work on one AMP than there is on the other AMPs.

For smaller tables, a larger skew factor doesn't have nearly the same impact.

Q – *How do Load Utilities deal with PPIs?*

A – There are no issues loading PPI tables using the utilities. We can use Teradata Parallel Transporter, FastLoad, MultiLoad, and TPump to load data into partitioned tables. Sometimes performance is better when loading into a partitioned table because we add data to only a partition.

Q – *If we have a Claim Table with a Partitioned Primary Index on ClaimId and it is partitioned on DATE (by month), can we add a specific date condition to the WHERE clause of our query?*

A - If we ask for a specific date in our query, the optimizer uses it to facilitate partition elimination.

If the query is based on a specific ClaimId, we look in every partition.

If we specify an exact date in the WHERE clause, we use partition elimination and reduce the disk I/Os.

If we specify a large range of dates, we look at all the partitions.

If we specify year, but not month, we only search the 12 partitions for that year.

Q – *How are copy/restores done with partitioned tables?*

A – There are a number of options with a partitioned table. We can archive an entire table or selected partitions of a table. As part of a restore or copy, we can restore selected partitions, or the entire table, assuming we archived the full table.

The archive facility can archive selected partitions and then copy a portion of the data in the table. Or we can archive the full table and restore selected partitions from it.

Q – *Does the total number of partitions allowed increase with Teradata Release 14?*

A – Yes. The number of PPI partitions is increased from 65,535 to 9.2 quintillion in Teradata Release14.

Q – *We want to use an Aggregate Join Index. One of the base tables gets dropped and recreated twice a day as part of a process. Do we have to drop and recreate the JI again, every time we drop and create a base table?*

A – Yes. We must drop and recreate the JI each time. In fact, we cannot drop a table while there a Join Index that references it.

Another solution is to "update" the base table. This avoids dropping and recreating the base table and the JI.

Q – *Is there a way to collect statistics on the multi-level partitions? Does it collect on all levels?*

A – The system does not collect statistics on the individual levels. The syntax is COLLECT STATISTICS ON <table> COLUMN PARTITION;. COLUMN PARTITION includes the combined partition number so it collects on the combination of the levels. Collecting on an individual level is not yet supported. For now, COLLECT STATISTICS is on the partitioning column.

Q – *PARTITION Statistics can be collected on PPI tables as well as Non-PPI tables and is said to be inexpensive. How does that work? Aren't PARTITION Statistics only for PPI tables?*

A – We can think of all Teradata tables as being partitioned. When the table does not have PPI defined, all the rows are in one partition. Therefore, PARTITION Statistics can be collected on Non-PPI tables, but it is the same as COLLECT STATISTICS.

Q – *What is the advantage of including the RowID in defining a Join Index over not including it?*

A – If all of the columns in the tables are contained in the Join Index, the Join Index can be used to cover the query and including the RowID doesn't make any difference. If any of the columns needed for the query are missing from the Join Index, then including the RowID can be useful to join back to the base table and access the necessary column(s). Consider including RowID in the Join Index, called a Global Join Index, where there is a possibility of joining back to the base table.

Q – *How can we determine if and when our indexes are being accessed?*

A – We've included the query in Figure 130 for you to use to determine when and how many times an index was last accessed. Just insert your database name where it says, "ReplaceWithYourDataBaseName".

```
SELECT      DatabaseName
            , TableName
            , IndexName
            , ColumnName
            , IndexType,
CASE        IndexType
            WHEN        'P' THEN 'Nonpartitioned Primary'
            WHEN        'S' THEN 'Secondary'
            WHEN        'K' THEN 'Primary Key'
            WHEN        'U' THEN 'Unique Constraint'
            WHEN        'Q' THEN 'Partitioned Primary'
            WHEN        'V' THEN 'Value Ordered Secondary'
            WHEN        'J' THEN 'Join Index'
            WHEN        'N' THEN 'Hash Index'
            WHEN        'O' THEN 'Value Ordered (All) covering secondary'
            WHEN        'H' THEN 'Hash Ordered (All) covering secondary'
            WHEN        'I' THEN 'Ordering column of composite secondary'
            WHEN        'M' THEN 'Multi-Column statistics'
            WHEN        'D' THEN 'Derived column partition statistics'
ELSE        IndexType
END         (TITLE 'Type')
,UniqueFlag AS "Unique"
,IndexNumber
,ColumnPosition
,AccessCount
,LastAccessTimeStamp AS "Last Access"
,CreatorName
,CreateTimeStamp
,LastAlterName
,LastAlterTimeStamp
FROM        DBC.IndicesV
WHERE       TRIM(ReplaceWithYourDataBaseName) LIKE 'PD'
AND IndexType NOT IN ('1','2','I','M','D')
ORDER BY DatabaseName, TableName, IndexNumber, ColumnPosition;
```

Figure 130: INDEX LastAccess and AccessCount

Figure 131 shows what the report looks like.

Answerset 1

		TableName	Index Name	ColumnName	Index Type	Type	Unique	IndexNumber	Column Position	Access Count	Last Access	Creator Name	CreateTimeStamp	LastAlterName	Last…
1	PD	Department	?	Dept_Number	P	Nonpartitioned Primary	Y	1	1	?	?	Sysdba	12/17/2010 22:14:17	Sysdba	12/…
2	PD	Department	?	Dept_Name	U	Unique Constraint	Y	4	1	?	?	Sysdba	12/17/2010 22:14:17	Sysdba	12/…
3	PD	Employee	?	Employee_Number	P	Nonpartitioned Primary	Y	1	1	4	5/18/2011 09:18:40	Sysdba	12/17/2010 22:14:17	Sysdba	12/…
4	PD	Emp_Phone	?	Employee_Number	P	Nonpartitioned Primary	N	1	1	?	?	Sysdba	12/17/2010 22:14:18	Sysdba	12/…
5	PD	GT_Deptsalary	?	Dept_Number	P	Nonpartitioned Primary	N	1	1	?	?	Sysdba	12/17/2010 22:14:18	Sysdba	12/…
6	PD	Job	?	Job_Code	P	Nonpartitioned Primary	Y	1	1	?	?	Sysdba	12/17/2010 22:14:18	Sysdba	12/…
7	PD	Job	?	Job_Desc	U	Unique Constraint	Y	4	1	?	?	Sysdba	12/17/2010 22:14:18	Sysdba	12/…

Figure 131: INDEX LastAccess and AccessCount Output

Row 3 represents the Employee Table with a UPI on Employee_ Number. We can see that the index was last accessed four times on May 18, 2011.

For more information about Alison Torres and Teradata Database Index Essentials, please visit:

http://TeradataIndexes.com

Index

A

Aggregate Join Index, xvii, 68, 123, 140

Access Module Processor, xvii, 2

AMP, xvii, 2, 3, 4, 6, 7, 8, 9, 10, 11, 13, 15, 16, 19, 20, 22, 23, 24, 29, 30, 35, 39, 47, 48, 50, 51, 53, 59, 61, 62, 63, 65, 68, 73, 81, 88, 91, 93, 94, 95, 96, 116, 117, 119, 125, 129, 131, 132, 133, 134, 137, 138, 139

C

Character PPI, 40, 44, 45

Compressed Single Table Join Index, 87

G

Global Join Index, 4, 69, 95, 113, 114, 115, 116, 117, 118, 119, 120, 121, 137, 141

H

Hash, 51

Hash Index, 4, 45, 115, 125, 126, 127

Hashed NUSI, 115

J

Join Index, xvii, 67, 68, 69, 70, 72, 73, 78, 80, 82, 83, 84, 85, 86, 88, 89, 90, 91, 92, 94, 95, 96, 97, 100, 101, 102, 103, 104, 105, 106, 108, 109, 110, 112, 113, 115, 118, 121, 122, 123, 127, 137, 140, 141

JI, xvii, 67, 68, 69, 73, 77, 80, 81, 84, 85, 91, 93, 94, 95, 96, 99, 102, 112, 123, 140

L

Local Hash, 51

M

Massively Paralleling Processing, xvii, 3

MPP, xvii, 3

Multi Level Partitioned Primary Index, xvii, 29, 31, 35, 38, 45

MLPPI, xvii

N

NoPI, 129, 130, 131, 132, 133, 134, 135

NPPI, xvii, 24

Non-Unique Primary Index, xvii, 5, 7, 13, 22, 25, 48, 137

NUPI, xvii, 7, 9, 51, 53, 59, 61, 62, 64, 85, 133, 134, 137

Non-Unique Secondary Index, xvii, 4, 6, 24, 47, 51, 52, 64, 90, 137

NUSI, xvii, 6, 24, 25, 47, 51, 52, 53, 54, 55, 56, 57, 58, 59, 62, 63, 64, 65, 90, 91, 94, 95, 96, 97, 105, 113, 115, 116, 117, 118, 123, 135, 137

NUSI Bit Mapping, 64

O

Optimizer, xvii, 6, 8, 17, 18, 21, 29, 50, 53, 54, 55, 58, 63, 64, 65, 67, 68, 69, 80, 81, 82, 84, 86, 89, 91, 96, 102, 121, 123, 125, 135, 137, 138, 139

P

Parallel Database Extensions, xvii, 1

Partitioned Join Index, 83, 121

PDE, xvii, 1

Parsing Engine, xvii, 2

PE, xvii, 2

Primary Indexes, xvii, 4, 5, 6, 9, 11, 47, 68, 91, 92, 95, 129

PI, xvii, 5, 8, 13, 14, 15, 17, 59, 61, 63, 67, 68, 85, 92, 93, 95, 123, 125, 127, 129, 131, 133, 134, 137, 139

Partitioned Primary Index, xvii, 4, 11, 12, 14, 16, 17, 20, 30, 107, 121, 137, 139

PPI, xvii, 11, 12, 13, 15, 16, 17, 22, 23, 24, 25, 29, 30, 45, 46, 59, 60, 61, 63, 64, 107, 108, 121, 139, 140, 141

S

Secondary Index, xvii, 4, 6, 14, 17, 25, 46, 47, 63, 64, 65, 67, 68, 69, 97, 113, 125, 127

Single Table Join Index, xvii, 4, 55, 68, 85, 87, 90, 91, 92, 94, 123, 125, 127

Sparse Partitioned Join Index, 112, 113

Sparse Join Index, 4, 55, 69, 99, 108, 110, 111, 112, 113, 123

STJI, xvii, 4, 55, 68, 87, 90, 91, 92, 93, 94, 95, 96, 97, 99, 102, 105, 107, 108, 125

Symmetric Multi-Processor, xvii, 2

SMP, xvii, 2

T

Teradata BYNET®, iii, xvii, 2, 3, 4, 10, 39, 48, 50, 51, 95

Time-Based PPI, 26, 27

U

Unique Primary Index, xvii, 5, 7, 9, 13, 18, 24, 35, 133, 137

UPI, xvii, 7, 9, 13, 51, 143

Unique Partitioned Primary Index, 35

Unique Secondary Index, xvii, 4, 7, 24, 47, 48, 49, 51, 59, 60, 137

USI, xvii, 7, 24, 47, 48, 49, 50, 51, 59, 60, 61, 64, 65, 91, 130

V

Value Ordered Join Index, 4, 14, 67, 97, 99, 137

Value Ordered NUSI, 57, 97, 99, 108

Value Ordered Sparse STJI, 4, 105, 107, 108